Advanced Sciences and Technologies for Security Applications

Indexed by SCOPUS

The series Advanced Sciences and Technologies for Security Applications comprises interdisciplinary research covering the theory, foundations and domain-specific topics pertaining to security. Publications within the series are peer-reviewed monographs and edited works in the areas of:

- biological and chemical threat recognition and detection (e.g., biosensors, aerosols, forensics)
- crisis and disaster management
- terrorism
- cyber security and secure information systems (e.g., encryption, optical and photonic systems)
- traditional and non-traditional security
- energy, food and resource security
- economic security and securitization (including associated infrastructures)
- transnational crime
- human security and health security
- social, political and psychological aspects of security
- recognition and identification (e.g., optical imaging, biometrics, authentication and verification)
- smart surveillance systems
- applications of theoretical frameworks and methodologies (e.g., grounded theory, complexity, network sciences, modelling and simulation)

Together, the high-quality contributions to this series provide a cross-disciplinary overview of forefront research endeavours aiming to make the world a safer place.

The editors encourage prospective authors to correspond with them in advance of submitting a manuscript. Submission of manuscripts should be made to the Editor-in-Chief or one of the Editors.

More information about this series at http://www.springer.com/series/5540

Margaret E. Kosal
Editor

Disruptive and Game Changing Technologies in Modern Warfare

Development, Use, and Proliferation

Springer

Editor
Margaret E. Kosal
Sam Nunn School of International Affairs
Georgia Institute of Technology
Atlanta, GA, USA

ISSN 1613-5113 ISSN 2363-9466 (electronic)
Advanced Sciences and Technologies for Security Applications
ISBN 978-3-030-28341-4 ISBN 978-3-030-28342-1 (eBook)
https://doi.org/10.1007/978-3-030-28342-1

This Springer imprint is published by the registered company Springer Nature Switzerland AG
The registered company address is: Gewerbestrasse 11, 6330 Cham, Switzerland

Acknowledgements

This work was initiated while the editor was serving as Director of the Sam Nunn Security Program (SNSP) at the Georgia Institute of Technology (aka Georgia Tech), where she is an Associate Professor of International Affairs. The mission of the SNSP is to strengthen and foster scholarship at the intersection of science, technology, and national and international security that address grand challenges of today and for tomorrow through the creation of educational opportunities for pre-doctoral to mid-career scientists and engineers, structured research projects, and promotion of opportunities for to inform and advise the policy community.

In addition to the contributing authors, the following persons are gratefully acknowledged for making this work a success:

- LTC Sean P. Feeley, US Army (ret.), Mr. Frank Reidy, Mr. Mark Raney, and everyone at Joint Special Operations University (JSOU), who supported and gave time to this project.
- Dean Jacqueline Royster, Dean of the Ivan Allen College of Liberal Arts at Georgia for her generous support of the SNSP.
- Ms. Heather Regnault, who served as editorial assistant on the project. While she is an independent scholar in her own right, I am ever grateful for her assistance.

Contents

Contributors

Andrew Conant Georgia Institute of Technology, Atlanta, GA, USA

Peter R. Exline Georgia Institute of Technology, Atlanta, GA, USA

Seth E. Gordon Georgia Institute of Technology, Atlanta, GA, USA

Margaret E. Kosal Georgia Institute of Technology, Atlanta, GA, USA

T. Christopher Petty US Army (Retired), Seattle, WA, USA

Heather Regnault Georgia Institute of Technology, Atlanta, GA, USA

Rockie Rodriguez Georgia Institute of Technology, Atlanta, GA, USA

Sara M. Schmitt Georgia Institute of Technology, Atlanta, GA, USA

Anmol Soni Georgia Institute of Technology & Georgia State University, Atlanta, GA, USA

Zoe Stanley-Lockman S. Rajaratnam School of International Studies, Singapore, Singapore

J. Wes Stayton Georgia Institute of Technology, Atlanta, GA, Georgia

Introduction

Margaret E. Kosal and Heather Regnault

Abstract This book explores and analyzes emerging innovations within today's most cutting-edge science and technology (S&T) areas, which are cited as carrying the potential to revolutionize governmental structures, economies, and international security. Some have argued that such technologies will yield doomsday scenarios and that military applications of such technologies have even greater potential than nuclear weapons to radically change the balance of power. As the United States looks to the future—whether dominated by extremist groups co-opting advanced weapons in the world of globalized non-state actors or states engaged in persistent regional conflicts in areas of strategic interest—new adversaries and new science and technology will emerge. Choices made today that affect science and technology will impact how ably the US can and will respond. Chapters within the book look at the changing strategic environment in which security operations are planned and conducted; how these impact science and technology policy choices made today; and predictions of how science and technology may play a beneficial or deleterious role in the future. Some game changing technologies have received global attention, while others may be less well known; the new technologies discussed within this proposal, as well as future discoveries, may significantly alter military capabilities and may generate new threats against military and civilian sectors.

Emerging innovations within today's most cutting-edge science and technology (S&T) areas are cited as carrying the potential to revolutionize governmental structures, economies, and life as we know it; others have argued that such technologies will yield doomsday scenarios and that military applications of such technologies have even greater potential than nuclear weapons to radically change the balance of power [1]. When asked what are the current approaches and thinking on means for deterring emerging technologies of concern (beyond cyber) to the U.S., then-US

M. E. Kosal (✉) · H. Regnault
Georgia Institute of Technology, Atlanta, GA 30332, USA
e-mail: margaret.kosal@inta.gatech.edu

H. Regnault
e-mail: hregnault3@gatech.edu

M. E. Kosal (ed.), *Disruptive and Game Changing Technologies in Modern Warfare*, Advanced Sciences and Technologies for Security Applications,
https://doi.org/10.1007/978-3-030-28342-1_1

Strategic Command (STRATCOM) Commander General Robert Kehler (USAF) responded that "surprise is what keeps me up at night" and cited current uncertainty in how to assess and address emerging and disruptive technologies [2]. These concepts and the underlying defense needs were articulated at the multi-national level in NATO's New Strategic Concept paper:

> Less predictable is the possibility that research breakthroughs will transform the technological battlefield. Allies and partners should be alert for potentially disruptive developments in such dynamic areas as information and communications technology, cognitive and biological sciences, robotics, and nanotechnology. The most destructive periods of history tend to be those when the means of aggression have gained the upper hand in the art of waging war [3].

While the suggestion that such emerging technologies will enable a new class of weapons that will alter the geopolitical landscape remains to be realized, a number of unresolved security puzzles underlying the emergence of potentially disruptive technologies have implications for international security, defense policy, governance, and arms control regimes. The extent to which these emerging technologies may exacerbate or mitigate the global security and governance challenges that states will pose in the future to U.S., regional, and global security interests will remain an integral question as US policymakers and leaders navigate the complex global environment.

For scholars of international security, the intersection of technology and understanding the outcomes of armed conflict has a rich literature [4–8]. Within science and technology studies, similarly there is a rich literature exploring the intersection of science, technology and military applications [9–11]. For strategists and scholars of revolution in military affairs (RMA) [12–17] and of fourth and fifth generation warfare [18–23], the nexus between technology and military affairs is not just speculation but a reality that bears directly on the propensity for conflict and outcomes of war, as well as the efficacy of security cooperation and coercive statecraft.

The research in this volume aims to address those security puzzles in order to understand and explain how emerging technologies affect strategic cooperation, conflict, and the future of security. This work explores via interdisciplinary methods how emerging technologies reconcile with or challenge traditional ideas for national security.

How, when, where, and in what form the shifting nature of technological progress may bring enhanced or entirely new capabilities, many of which are no longer the exclusive domain of a single nation-state, is contested and requires more cross-disciplinary thinking. Contemporary analyses of these emerging technologies often expose the tenuous links or disconnections among mainstream scholarship on international security, understanding of the military technological innovation and acquisition processes, disconnections among the scientific and technical realities and mainstream scholarship on national and international security, especially with regard to potential to have impact on strategic and policy. Conceptually, technologies can be seen as evolutionarily advancing current capabilities or pressing to the bleeding edge and enabling disruptive, revolutionary capabilities developments. The ability to differentiate or gain insight into such has thus far not been explored or analyzed robustly with respect to strategic implications beyond a technologically-deterministic lens.

The novel scientific principles that underlie the character of these uncertain technologies and their convergence with political and social institutions reveal conceptual and empirical confusion associated with assessing the national security implications. There also is palpable confusion over the technical and strategic distinguishability and dominance of prospective offensive and defensive systems. As an example, while the suggestion that nanotechnology will enable a new class of weapons that will alter the geopolitical landscape remains to be realized, a number of unresolved security puzzles underlying the emergence of potentially game-changing technologies have implications for international security, defense policy, governance, and arms control regimes [24].

In the post-Cold War environment, the most technologically advanced military power no longer guarantees national security. Globalization and the information revolution, including the Internet and other communication leaps—have led to much greater visibility into the availability and potential for technology [25]. New technological developments have become accessible and relatively inexpensive to a larger number of nations and within the grasp of non-state actors: advanced technology is no longer the domain of the few [26]. In the 21st century, both nation-states and non-state actors may have access to new and potentially devastating dual-use technology [27, 28]. Additive manufacturing, artificial intelligence, and CRISPR technology are examples of such emerging technology that have dual use applications. For this volume, the dual-use and the dual-use conundrum refers to the fact that almost all the equipment and materials needed to develop dangerous or offensive agents, particularly biological and chemical agents, have legitimate uses in a wide range of scientific research and industrial activity, including defensive military uses. Within this text it does not refer to the demarcation between civilian and military uses. Understanding these changing paradigms and limiting the proliferation of unconventional weapons for the 21st Century starts with an awareness of the factors driving the capabilities, understanding the underlying science and the challenges of defense, considering the changing nature of technological progress and the changing nature of warfare, and the relationship between science and security domestically and internationally.

Communication of those new discoveries is occurring faster than ever, meaning that the unique ownership of a piece of new technology is no longer a sufficient position, if not impossible. It is widely regarded that recognition of the potential applications of a technology and a sense of purpose in exploiting it are far more important than simply having access to it today [29]. Technological surprise has and will continue to take many forms. A plethora of new technologies that are under development for peaceful means but may have unintended security consequences and will certainly require innovative countermeasures.

Reducing the risk from misuse of technology will mean consideration of the highly transnational nature of the critical technology required. Traditional and innovative new approaches to nonproliferation and counterproliferation are important policy elements to reduce the risk of malfeasant application of technology that may enable advanced weapons or make production or dissemination of biochemical agents available to a much wider group of actors. Efforts to strengthen existing international regimes to control transfers of dual-use materials are ongoing [30]. Verification still remains a technical as well as diplomatic challenge. The role of international agreements and cooperative programs in the 21st Century is a contested intellectual and policy field.

As nations and the international community look to the future—whether dominated by extremist groups co-opting advanced weapons in the world of globalized non-state actors or states engaged in persistent regional conflicts in areas of strategic interest—new adversaries and new science and technology will emerge. Choices made today that affect science and technology will impact how ably the US can and will respond. The changing strategic environment in which security operations are planned and conducted impacts science and technology policy choices made today and affects how science and technology may play a beneficial or deleterious role in the future. Some game-changing technologies have received global attention, while others may be less well known; these new technologies and discoveries may significantly alter military capabilities and may generate new threats against military and civilian sectors.

Future trend analysis is a tricky task. Colin Gray said, "Trend spotting is easy. It is the guessing as to the probable meaning and especially the consequences of trends that is the real challenge." [31]. How, when, where, and in what form the shifting nature of technological progress may bring enhanced or entirely new capabilities, many of which are no longer the exclusive domain of the United States, is contested and requires better analytical tools and assessment to enable strategic responses. Contemporary analyses of these emerging technologies often expose the tenuous links or disconnections among mainstream scholarship on international security, understanding of the military technological innovation and acquisition processes, disconnections among the scientific and technical realities and mainstream scholarship on national and international security, especially with regard to potential to have impact on strategic and policy. Conceptually, technologies can be seen as evolutionarily advancing current capabilities or pressing to the bleeding edge and enabling disruptive, revolutionary capabilities developments. The ability to differentiate or gain insight into such has thus far not been explored or analyzed robustly with respect to strategic implications beyond a technologically-deterministic lens.

The novel scientific principles that underlie the character of these uncertain technologies and their convergence with political and social institutions reveal conceptual and empirical confusion associated with assessing the national security implications. While the suggestion that certain emerging technologies, such as additive manufacturing and biotechnologically-enabled weapons, will enable a new class of weapons that will alter the geopolitical landscape remains to be realized, a number of unresolved security puzzles underlying the emergence of such technologies have implications for international security, defense policy, governance, and arms control regimes. The research contained in this volume advances the strategic understanding of these game-changing technologies and the development of meaningful and testable metrics and models to help reduce that surprise.

The development, use, and proliferation of game-changing and disruptive technology are analyzed in this volume with respect to modern warfare and are organized into three main themes: (I) the adoption, proliferation, and governance of disruptive technology; (II) challenges of disruptive technologies to strategic stability; and (III) the effects of disruptive technologies on military capabilities and operations.

1 Adoption, Proliferation, and Governance of Disruptive Technology

Technology with the potential for disruptive applications must first be adopted, and if widespread adoption results, international governance is often viewed as essential in establishing norms of use and nonproliferation standards. In many cases technological advances either never make it to market or are otherwise limited in broad adoption. When a disruptive technology, such as the internet, does gain broad adoption and use, however, international norms and standards that broadly govern the technology may be slow to develop as international actors vie to lead in shaping international standards that benefit their own objectives in use. Further, the use of military technology adopted by nation states may be subject to international rules of war and further limited by constraints arising from ethical considerations in their use and interoperability concerns. As emerging technology rapidly advances, considerations of which countries will adopt these technologies, how the technologies may affect proliferation of WMD, and future issues related to governance are essential.

Proliferation of these game-changing technologies also affects the US' ability to maintain a technological edge against near-peer adversaries, as is currently the case with stealth and precision guided weapons technologies. Chapter 2 focuses on factors that influence technology adoption and governance of such technology using a framework that assesses a society's acceptance of a novel technology within the context of Revolutions in Military Affairs (RMA). As the US is currently exploring disruptive technologies that will enable a third offset in maintaining a military edge over rival nations, a discussion of potential influences on adoption and governance of emerging technologies is important in guiding the US military within this process as well as US national security policy. The author concludes that treaties and bans on technology have historically not managed the proliferation and use of military technology, though some treaties, such as the nuclear nonproliferation treaty, have had more impact than others.

As an example of how ethical considerations and political issues may influence the development, proliferation and use of technology, Chap. 6 analyzes the debate concerning the use of directed energy non-lethal weapons systems. As non-governmental organizations such as the International Committee of the Red Cross (ICRC), have advocated against the use of this technology, an analysis of the potential benefits versus the costs of utilizing such weapons systems is a useful place to start parsing the effects of ethical considerations and the laws of armed conflict, such as the Geneva Convention, on the use of emerging technology. The author assesses the utility of directed energy non-lethal weapons, defined as weapons that use electromagnetic or acoustic means to affect a target, by comparing their use to conventional weapons in three classes: unguided weapons, guided and unpowered weapons, and those which are guided and powered. After analyzing the utility of directed energy non-lethal weapons systems as well as the arguments against their use by opponents of their use, the author concludes that their utility outweighs the ethical arguments advanced in opposition to these weapon systems.

Once disruptive technologies are developed, the question of proliferation becomes salient. Chapter 9 attempts to develop a framework to predict proliferation of disruptive technologies by examining lethal autonomous weapons systems, defined as systems that operate without human intervention after their initial deployment. Given the investments by the US, Russia, and China in artificial intelligence and robotics, research and development in lethal autonomous weapons is expected to continue to expand rapidly. Drones are now used frequently by numerous nation states, though the US maintains the clear advantage in weaponized drones. Utilizing a framework developed in the nuclear nonproliferation community, the author examines the characteristics of countries that currently possess lethal autonomous weapons systems in an effort to determine the potential utility of the framework in predicting countries that may proliferate such systems and alter the balance of power in the global system. The author concludes that possession of nuclear weapons, research and development spending, and military spending are all factors that indicate a potential for a country to develop lethal autonomous weapons systems, though the model tested using quantitative methods does not have a predictive capacity.

2 Challenges to Strategic Stability

Researchers and analysts have suggested that disruptive technologies may ultimately challenge strategic stability as more countries develop advanced technology such as biotechnologically-enabled weapons, artificial intelligence, and machine learning. As nuclear weapons states' compete in a technological race to develop these weapons systems, the question of how these weapons may alter or even fundamentally change the nuclear weapons balance emerges. Further, the question of how emerging technology with dual use applications may affect nuclear nonproliferation and the ease with which non-state actors might obtain weapons of mass destruction are important aspects of the emergence of these disruptive technologies. As more nations compete to gain a competitive advantage over their adversaries via technological advances, the security dilemma, known as the effect that increasing one's own capabilities increases adversaries' uncertainty that leads them to increase their own capabilities, becomes more prominent in the context of emerging technology.

The potential for biotechnology to change the nature of war, whether by weapons designed with embedded technology or enhancements to military personnel have been noted by scientists and the US government alike [32]. Chapter 3 addresses the potential for biotechnology to alter the nature of strategic stability while discussing the implications of these technological advances in the context of traditional models of strategic stability. CRISPR, a gene-editing tool currently being researched extensively in China, is one such technology that may ultimately compete with nuclear weapons in the global strategic environment. The author offers an analysis of this technology with respect to global strategic stability as US policymakers must anticipate the emerging technologies that may become realized and potentially affect the US doctrine on nuclear deterrence.

As emerging technologies become more advanced and dispersed in the international global system, the question of their impact on global stability must be understood within the characteristics of the current international system. Disruptive technologies of war may impact the security dilemma by altering the perceived offense-defense balance, thereby altering global strategic stability. Chapter 8 analyzes metamaterial adaptive camouflage as an example of anti-detection technology, which reduces information available to adversaries, and potentially destabilizes the global environment as the author argues that this renders the technology as favoring the offense. By analyzing the historical aspects of detection technology within the context of disrupting strategic stability, the authors assesses whether anti-detection technology has been disruptive to the stability of the international system. As stealth technology has advanced historically, counter-technology advances as well, negating the disruptive effects; however, metamaterials in the hands of non-state actors could be quite disruptive to the international strategic stability.

Further implications of emerging technology include the potential to undermine the nuclear nonproliferation regime and contribute to the proliferation of other weapons of mass disruption (WMD). One such technology, additive manufacturing, also known as 3-D printing, has commercial and recreational applications with the potential to transform capabilities of nation-states and non-state actors. As with other emerging technologies, the internet contributes to the potentially revolutionary nature of additive manufacturing technology with print files readily available on the internet for individuals who seek to use the technology. The availability for individuals to obtain print-files for handguns or other weapons causes concern that this technology may undermine international regulations and regimes that attempt to limit their proliferation. Chapter 4 discusses the risk of the dual-nature of additive manufacturing to undermine the nuclear nonproliferation regime as the technology has the potential to reduce the cost and availability of manufactured parts for use in weapons systems and ease the transfer of sensitive nuclear technologies. Further, the reduction in facility size that 3-D manufacturing technology offers could decrease the US' and international agencies' ability to detect potential proliferation threats. The author concludes that additive manufacturing, in the technology's current state, does not represent a high risk to nuclear proliferation in the areas of nuclear enrichment and reprocessing, but represents a moderate risk in some chemical weapons applications. However, the risk with respect to WMD delivery mechanisms is considered higher and the author suggests the US monitor advances in the field as a more mature market could change the risk of this technology with respect to WMD and proliferation.

Whereas emerging technology has the potential to undermine proliferation of WMD, certain technologies may also have the potential to strengthen counter WMD efforts. With the advent of machine learning and artificial intelligence, intelligence indicators of potential proliferation may be enhanced and the US' ability to monitor potential states' WMD proliferation facilitated. Machine learning utilizes advances in computational power to analyze large amounts of data in a relatively short period of time. Chapter 5 develops a framework, using six key intelligence indicators, for assessing the feasibility and utility of applying machine learning, specifically reinforcement learning, to US efforts to counter WMD proliferation. The author analyzes the effectiveness of machine learning as compared to a human operator in assessing intelligence about the manufacture, weaponization, obtainment of fissile material and uranium and plutonium enrichment beyond a certain level, all of which are indicators of nuclear weapons proliferation. The author further examines the utility of machine learning in assessing potential indicators of proliferation of chemical and biological weapons such as hidden facilities, development of delivery systems, and purchases of protective gear. Limitations in machine learning are assessed with respect to factors such as data set size, which can limit the machine's ability to learn, uncertainty, and near future potential in nation-states which may potentially proliferate WMD. The author concludes that assessment of indicators such as excess power generation in an area without a clear need (suggesting a covert enrichment facility), evidence of large scale excavations indicating covert facilities, and communications metadata analysis are areas in which machine learning could be most useful to US counter WMD efforts. Finally, the analysis suggests that future use of generalized AI in counter WMD efforts is likely to be fruitful.

3 Effect on Military Capabilities and Operations

After the Revolution in Military Affairs (RMA) of the 1970s whereby the US introduction of precision-guided weapons systems changed the operational nature of warfare, the US displayed the integration of this and other technologies into a new war-fighting operational concept in the 1991 Gulf War. After countries, such as Russia and China, analyzed the advancement in US military technology and operational concepts, they began developing plans to invest in their own military modernization plans. As these near-peer competitors close the gap in the US military technological prominence, the US is developing plans for a third offset, an effort to widen this gap by investing in emerging technology, restructuring the military procurement process and establishing a more efficient and effective fighting force given the reemergence of great power competition after the end of the Cold War. The technological arms race that is ensuing is a reflection of these factors.

The potential for disruptive technologies to wreak havoc on US military endeavors has already been noted in areas such as GPS-manipulation and directed energy weapons. In a global environment in which near-peer competitors are engaging in gray zone tactics, below the threshold which they anticipate a US military response, the US must anticipate potential disruptive uses of emerging technology on US military operations. A forward-looking analysis on such disruptive technologies is essential for the US to anticipate these disruptions and prepare alternative military operational concepts to maintain resilience in a technologically-enabled warzone. The potential for disruptive technologies to advance the US' objectives of maintaining military dominance over near-peers as well as the potential threats to US military performance initiated by their emergence is an essential area to be investigated.

The US' ability to maintain its technological superiority and establish operational methods that incorporate emerging technology while changing the nature of warfare is essential to US deterrence and effectiveness of potential future military engagements with near-peer adversaries and non-state actors. In an age where emerging technologies are being rapidly developed, appropriate US military investment decisions in emerging technology are crucial. Estimating the trajectory of emerging technologies in terms of future utility, viability, effectiveness, and potential for incorporation into existing military infrastructure is a complex one, but nonetheless an important endeavor. Referencing the Department of Defense's "Defense Innovation Initiative," Chap. 10 provides an assessment of future investment areas for the US Army to maintain its military edge over adversaries, such as Russia and China. The author examines the potential for emerging technologies to enhance US Army Aviation platforms that would initiate an asymmetric advantage for the US in fighting near-peer and non-state actors. The author performs a comprehensive analysis of the potential utility of disruptive technologies in advancing Future Vertical Lift Advanced Protection Systems and an evaluation of the potential future operating conditions in which the US Army will fight. Based upon this analysis, the author recommends the US Army invest in nanotechnology, artificial intelligence, and quantum computing to enable Future Vertical Lift and develop the operational concepts to facilitate the third offset.

As the US military seeks to close the gap in technological advantages between the US and near-peers, an important area within a third offset strategy lays in operational and organizational aspects of the US military, including military logistics. The US military acknowledges the role of military logistics, along with technological advances and changes in military strategic and operational concepts, in initiating a Revolution in Military Affairs. With fewer forces in Europe and Russia's emphasis on swift military movements, prioritizing advances in military logistics is essential in reestablishing and maintaining US military dominance. Chapter 11 builds upon the US Army's 1999 Revolution in Military Logistics (RML) concept by assessing the potential of emerging technologies, such as artificial intelligence, alternative energy technologies, augmented reality, additive manufacturing, and the internet-of-things to facilitate a RML in the future. Acknowledging the connection between military strategy and logistics, the author assesses six tenets underlying a RML, each relying on information technology for its performance, and argues that the technological piece of a RML has thus far been applied in an ad hoc manner. Additionally, the implementation of technology in logistics will further require changes in training for logisticians, which should be incorporated in the US military's plans for modification of the logistics element of military operations. Although the chapter concludes that technology can enhance the transformation of US military logistics, the task is complex, and the emerging technologies also present risks of their own, including the rapid advancement of such technologies and the technical skills required for their use.

The US military is also facing pressure to modify its strategy with respect to energy efficiency and energy sources as energy requirements are expected to increase due to technologically-based operations. As such, an analysis of potential advancements in energy technology that could potentially transform military operations is essential for military leaders to plan and implement strategies that maintain the military's operational effectiveness and resilience. Chapter 7 analyzes potential energy-related technological advancements in overseas US military operations and assesses the utility of a framework to evaluate the extent to which these technologies could be considered game-changers. The author assesses these technologies within the energy value chain, including generation, storage, transfer and use, and the extent they could become game-changers for the US military as assessed using the Lethality-Survivability-Mobility framework. Individual applications of energy technology are recommended for further research and development as they may have the most potential to positively affect mobility and survivability. Recommendations for the US military to cooperate with the private sector in customizing and integrating existing energy technology into US military operations are also outlined.

The current explosion in emerging technologies offers great opportunity for the US military to maintain its competitive technological and operational advantage over adversaries. At the same time, these potentially disruptive technologies create challenges for the US and for the international community. Development of norms of use with respect to emerging technologies typically lag behind their rapid advancement while international actors vie to sustain or develop a competitive edge may view norms of use as secondary to their national security objectives. Proliferation of these technologically advanced weapons systems also poses a concern given the potential for some to be used as WMD, while the technologies may also enhance the US' ability to detect proliferation in WMD. Whether and how emerging technologies will alter the global strategic balance remains to be seen, but the US should continue to evaluate the potentially disruptive effects of rapidly emerging technology. This book aims to contribute to this type of evaluation through qualitative and quantitative analyses of the potential development, proliferation and use of game-changing technologies in modern warfare.

References

1. Jeremiah DE (1995) Nanotechnology and global security. Palo Alto, CA, Fourth foresight conference on molecular nanotechnology, 9 November 1995
2. Sustaining the Triad: the Enduring Requirements of Deterrence" Conference, 8 November 2013, Naval Submarine Base Kings Bay, Georgia
3. Active Engagement, Modern Defence: Strategic Concept for the Defence and Security of the Members of the North Atlantic Treaty Organization. Adopted by the Heads of State and Government at the NATO Summit in Lisbon, 19–20 Nov 2010
4. Rosen SP (1991) Winning the next war: innovation and the modern military. Cornell U Press
5. Skolinikoff EB (1993) The elusive transformation: science, technology, and the evolution of international politics. Princeton U Press

6. Solingen E (1994) Scientists and the state: domestic structures and the international context. U of Mich Press
7. Cronin PM (ed) (2008) Impenetrable fog of war: reflections on modern warfare and strategic surprise. Praeger Security International
8. O'Hanlon ME (2009) The science of war. Princeton U Press
9. Long FA, Reppy J (eds) (1980) The genesis of new weapons: decision making for military R&D. Pergamon Press
10. Gusterson H (2005) A pedagogy of diminishing returns: scientific intuition across three generations of nuclear weapons science. In: Pedagogy and the practice of science. MIT Press, pp 75–107
11. Hacker BC, Hacker M (2006) American military technology: the life story of technology. Johns Hopkins Press
12. Blank SJ (1984) The Soviet strategic view: Ogarkov on the revolution in military technology. Strateg Rev 12:3–90
13. Krepinevich AF (1994) Cavalry to computer: the pattern of military revolutions. Natl Interest 37:30–42
14. Nye JS Jr, Owens WA (1996) America's information edge. Foreign Aff 75:20–36
15. Cohen EA (1996) A revolution in warfare. Foreign Aff 75:37–54
16. Bernstein AH, Libicki M (1998) High-tech: the future face of war? A debate. Commentary 105:28–31
17. Arquilla J (2003) Worst enemy: the reluctant transformation of the American military. Ivan R Dee, Lanham, MD
18. Lind WS, Theile G (2015) 4th generation warfare handbook. Castalia House
19. Lind WS (2004) Understanding fourth generation war. Mil Rev 84:12–16
20. Hammes TX (2006) The sling and the stone: on war in the 21st century. Zenith Press
21. Benbow T (2008) Talking 'bout our generation? Assessing the concept of 'fourth-generation warfare'. Comp Strategy 27:148–163
22. Artellia MJ, Deckrob RF (2008) Fourth generation operations: principles for the 'long war'. Small Wars Insur 19:221–237
23. McKenzie JF Jr (1993) Elegant irrelevance: fourth generation warfare. Parameters, pp 51–60
24. Kosal ME (2009) Nanotechnology for chemical and biological defense. Springer, New York
25. Rennstich JK (2008) The making of a digital world: the evolution of technological change and how it shaped our world. Palgrave MacMillan, New York
26. Office of the Director of National Intelligence (2007) Unclassified Key Judgments of the National Intelligence Estimate, 'Prospects for Iraq's Stability: A Challenging Road Ahead.' 2 Feb 2007
27. National Research Council (2006) National Academy of Sciences, Biotechnology Research in an Age of Terrorism, National Academies Press: Washington, DC
28. National Research Council (2004) National Academy of Sciences, Globalization, Biosecurity, and the Future of the Life Sciences. National Academies Press: Washington, DC
29. Krulak CC (1999) USMC (ret), The strategic corporal: leadership in the three block war. Marines Magazine, January 1999. http://www.au.af.mil/au/awc/awcgate/usmc/strategic_corporal.htm
30. Kosal ME (2008) U.S. policies to reduce the threat of chemical terrorism. In: Prepared for the partnership for a secure America, 9/11+6 initiative foreign policy priorities for a secure America, May 2008
31. Gray Colin (2007) Another bloody century: future warfare. Phoenix, London, UK, p 38
32. Kott A et al (2018) Potential science and technology game changers for the ground warfare of 2050: selected projections made in 2017. US Army Research Laboratory, February 2018. https://www.arl.army.mil/arlreports/2018/ARL-TR-8283.pdf

Game-Changing Military Technologies: Adoption and Governance

Rockie Rodriguez

Abstract This chapter offers an analysis and discussion of disruptive technology and revolutions in military affairs. It explores factors that influence technology adoption and analyzes whether certain legal and diplomatic responses (e.g., treaties, bans, and prohibitions) to weapons innovations are an effective governance tool for limiting proliferation. Examining the outcome of governance tools on the adoption and proliferation of technology can elucidate trends, uncover relationships, and inform better approaches to international security, diplomacy, and policy. Knowing what relationships exist is important to developing a clear picture of viable options and possible leverage points. Investigating the interaction between technology and governance enhances the knowledge base from which national security decisions are made and may contribute to increased effectiveness in governance approaches. Making the best, most inform decisions, is an imperative aspect in maintaining superiority.

1 Introduction

One way countries seek to increase military advantage is through the research and development of new technologies. Many advancements in capabilities are incremental, but at times a new game changing technology emerges that has the potential to disrupt the existing governance system and the established rules and traditions of warfare. This is called a disruptive or game changing technology. While a new and effective means of warfare is a goal of weapons development and considered an innovate success, game changing technologies are also a threat to international players and to global stability. At times, governance systems attempt to ban the disruptive technologies but there exists an uncertainty as to whether these bans are effective. Fully understanding the trajectory of technology adoption and the factors that influence technology proliferation is important to developing an effective national security strategy.

R. Rodriguez (✉)
Georgia Institute of Technology, Atlanta, GA, USA
e-mail: Rockie@gatech.edu; RockieRod@yahoo.com

M. E. Kosal (ed.), *Disruptive and Game Changing Technologies in Modern Warfare*, Advanced Sciences and Technologies for Security Applications,
https://doi.org/10.1007/978-3-030-28342-1_2

Because the banning of disruptive weapons is a repeated pattern in international affairs, it is worth consideration. Exploring the issues through an international relations lens, this chapter explores how treaties and bans temper or enhance technology adoption and weapons proliferation. The chapter starts by providing a discussion on the definition of disruptive technology and sets disruptive technology within the context of revolutions in military affairs and military offsets strategy. Using a framework proposed for determining which innovations will be accepted by society, this chapter then analyzes the adoption of various military technologies and maps to them bans and/or treaties to limit their use and proliferation. The final section is a discussion and a conclusion. In addition to the analysis, this chapter provides a foundation from which to engage in further discussions and expand research in military affairs and diplomacy.

2 What Is a Game Changing or Disruptive Technology? Developing a Definition

The definition of game changing or disruptive technology is critical to distinguishing it from standard technological advances. Joseph L. Bower and Clayton M. Christensen are credited [1] for coining the term disruptive technology in their 1995 publication in Harvard Business Review, *Disruptive Technologies: Catching the Wave* [2]. Bower and Christensen identify two classes of technology. Although originating from a description of technology in a business environment, term is also used to describe technology in military affairs. For Bower and Christensen, technology is either sustaining or disruptive. Sustaining technology has more or better attributes than that which is already valued. Sustaining technology is an incremental advance. An example of sustaining technology, as described by Bower and Christensen, is the replacement of ferrite heads and oxide disks by thin film disks for data storage. The change in storage capability of a particular hardware was a better storage technology than what already existed but it maintained the trajectory of performance.

The other type of technology, as coined by Bower and Christensen, is disruptive technology. Disruptive technology contrasts with sustaining technology and is described as a technology that introduces attributes that are very different from the ones historically valued. The decreasing size of the computer disk drives is described by Bower and Christensen as a disruptive technology. The decreasing size of the computer architecture created attributes—smaller size, internal power supplies, lighter-weight—that permitted large and bulky computers to develop into desktop PCs and laptops. This shift to smaller and mini computers is disruptive.

In a 2015 article [3], Christensen and colleagues revisit the term disruptive technology twenty years after its debut. They emphasize the distinction between sustaining and disruptive technology and caution against conflating the two: "We can't manage innovation effectively if we don't grasp its true nature." [2]. Managing technology, whether through treaties and bans or other governance approaches, is one of the themes of this chapter.

Though they use the term, not all researchers use a definition of disruptive technology that is consistent with what Bower and Christiansen have outlined. In 2010, the National Research Council (NRC) released a report, *Persistent Forecasting of Disruptive Technologies.* This report by the NRC is one of two reports funded by Department of Defense to provide an analysis of disruptive technologies. The report defines disruptive technology as "an innovative technology that triggers sudden and unexpected effects." [1]. The NRC acknowledges Bower and Christensen's formative work and describes the same two classes of technology but labels them as both having the potential to be disruptive. In the report, the NRC identifies two classes of disruptive technologies. The first class of disruptive technology replaces an existing technology by an incremental development. DVDs replacing VHS tapes and automobiles replacing horse and buggies are examples of incremental developments in the first class of disruptive technology. This is roughly analogous to the sustaining technology Bower and Christensen describe. The second class of disruptive technology identified by the NRC creates a new, market, or capability where none existed before. This is what is referred to as "novel capability" in Chart 1. Personal computers (PC), the internet, and stealth technology fall in this category. The second class of novel technology is what is described by Bower and Christensen as disruptive (Table 2.1).

Using similar definitions and examples, Bower and Christensen identify as a sustaining technology the same things that the NRC describes as an incremental development. While Bower and Christensen contrast sustaining and disruptive technology, the NRC organizes incremental development as a class of disruptive technology. The two camps agree that a disruptive technology delivers a new market, capability, or attribute, where none existed before.

Table 2.1 Classes of disruptive technology

Source	Class 1	Class 2	Notes
National Research Council	Novel capability—a class of technology that creates a new capability where none existed before	Incremental development—a class of technology replaces an existing technology by an incremental development	Novel capability and incremental developments in technology are different classes of the same
Bower and Christensen	Disruptive technology—introduces attributes that are very different	Sustaining technology—maintains the trajectory and improves the performance of a technology. More and better of the same technology	Disruptive and sustaining technology are inherently different

To be disruptive, technologies do need not be radical or novel from an engineering or technical perspective. In fact, another class of disruptive technology is important to acknowledge: Innovative use of existing technology. Using a combination of existing technologies in ways that are novel can result in a capability that is disruptive. The World Wide Web converging with the computer, communications, and browser technologies gave rise to the new information age and is one example of how converging, or synergistic, use of technology can be disruptive. Facebook is another example of how novel use of existing technologies can be disruptive. Neither the internet nor social networking were new in 2004, but Facebook brought the internet and networking together in an innovative way that was adopted by society. Facebook, as a social networking technology, has been a disruptive technology that changed the way people in many parts of the world relate to one another, interact on a daily basis, and receive news. Business models and governance strategies have also been revised to account for Facebook market.

Disruptive technology is distinctive because it upsets the established way of doing things. Disruptive technology causes shifts that change the world. Novel technologies are one of the principal means of surprising advisories or competitors and of disrupting established ways of doing things. It is, however, important to recognize that not all innovative, novel, new or emerging technologies or innovative use of technology are disruptive. Some new technologies and capabilities stay in the laboratory, many start-ups fail when taking the technology to market, and plenty of new and innovative technologies or use of technology never disseminate. Facebook, for example, was not the first social networking internet site, but it gained a broader acceptance than its predecessors.

When examining a potentially disruptive technology, the scale of dissemination is a useful factor in determining whether a technology is truly disruptive. Adoption is one critical measure of a technology becoming a disruptive technology. If a technology is not adopted, then it cannot be employed. Understanding what technologies are adopted and then disseminated widely is key to determining which technologies will earn disruptive status.

Based on the discussion and sources above, for the purposes of this chapter, disruptive technology is defined as:

> An innovative technology or use of technology that triggers 1) unexpected effects and 2) upsets the established way of doing things.

This definition of disruptive technology draws on various sources and provides a distinction from "normal" technology because of the scale of its impact. As discussed above, not all scholars agree on the criteria for disruptive technology. What is important to garner from this definition is that disruptive technology has a wide and profound impact on the established ways of doing things. By very nature, global stability can be challenged by technology that disrupts the established governance system.

3 Revolutions in Military Affairs

Countries seek to increase military advantage through the research and development of new technologies. Not all new technologies are game changing disrupters. Some new technologies are incremental or sustaining advancements, part of the evolution of technology. But other technologies are true game changers. They are disruptive. They are revolutionary. Revolutions in military affairs (RMA) are based on game changing technology that lead to changes in the character and conduct of conflict.

A revolution in military affairs can be characterized when:

> The application of new technologies into a significant number of military systems combines with innovative operational concepts and organizational adaptation in a way that fundamentally alters the character and conduct of conflict. It does so by producing a dramatic increase-often an order of magnitude or greater-in the combat potential and military effectiveness of armed forces. [4]

The important point in this definition is that technology's role is fundamental in a revolution in military affairs, but technology is also part of a larger phenomenon. According to scholars in the field [5], a revolution in military affairs is concept that requires supportive change in four areas: (1) technology, (2) systems development, (3) operational innovation, and (4) organizational adaptation. While each of the four elements is necessary for a revolution in military affairs, this chapter focuses on the technology aspect.

According to some scholars [4, 6], ten RMAs have taken place: (1) the Infantry Revolution, (2) the Artillery Revolution, (3) the Revolution of Sail and Shot, (4) the Fortress Revolution, (5) the Gunpowder Revolution, (6) the Napoleonic Revolution, (7) the Land Warfare Revolution, (8) the Naval Revolution, (9) the Mechanical Revolution, and (10) the Nuclear Revolution. Each revolution had a technology or group of technologies that supported, or in some cases, led it. Each revolution led to changes in the character and conduct of conflict, and each revolution has a technology or group of technologies that is credited for leading the revolution.

Here again, not all scholars agree. Alvin and Heidi Toffler offer an alternative view of RMAs [7]. In the book, *War and Anti-War: Survival at the Dawn of the 21st Century*, the Tofflers assert that only three military revolutions have occurred: (1) the agricultural revolution of 10,000 year ago, (2) the industrial revolution of 300 years ago, and (3) and the current revolution that is presently underway. Embedding military revolutions into the structure of the agricultural revolution, the industrial revolution, and the current (unnamed) revolution, they only consider a true revolution as one that "change[s] the game itself, including its rules, its equipment, the size and organization of the 'team,' their training, doctrine, tactics, and just about everything else." The agricultural revolution and the industrial revolution changed the way humans lived and worked, changed the way wealth was created, and changed the way wars were fought. Explanations of technological revolutions in war do not consider sustaining technologies, new uses of technology, or new combinations of existing technologies as disruptive or revolutionary. In the Tofflers' view the crossbow, gun powder, or nuclear weapons are not a significant part of an RMA.

While important for the reader to have an awareness of the differing views of RMA, this broad definition of revolutions does not provide granularity needed for purposes of this chapter and it has limited ability to translate to meaningful policy.

It is commonly accepted that we are currently in a revolution in military affairs, even Toffler and Toffler agree to this [8]. New technologies are supporting new military systems and operations integrations. Stealth weapons, precision weapons, smart weapons systems, information dominance, computers, autonomous weapons, and cyber technology, are some of the technologies supporting the current revolution in military affairs [5]. Experts are still trying to clearly define the current RMA and pin-point the technologies that will lead and support it. The major attempt to lead the current RMA comes in the form of the Third Offset Strategy.

4 Offset Strategies

An offset strategy is a central concept applied to the US military. Offset strategies have used technical innovation to counter the strength of military adversaries. There have been three offset strategies since they began in the 1950s.

The first offset strategy used a nuclear based deterrence strategy to offset Soviet land forces, proximity to Europe, and conventional superiority in Europe. In order to counter and deter the Soviet adversary, the US relied on massive retaliation and use of nuclear weapons. The first offset strategy was a success.

The second offset began 1970s. As the USSR developed their nuclear arsenal and delivery systems, a new strategy was needed to counter and deter the Warsaw Pact's numerically superior conventional forces and address Soviet advances in strategic nuclear capabilities in the late stages of the Cold War. The second offset strategy invested in the development of stealth aircraft, precision guided munitions, and space-based reconnaissance and navigation capabilities. Second offset capabilities and US military superiority were demonstrated during the First Gulf War.

Disruptive technology of the second offset has proliferated widely and adversaries (specifically, near-peer competitive states) have narrowed the technology gap. In 2014, the call for the game changing third offset was announced publically by then Secretary of Defense Chuck Hagel [9]. This details were further articulated by Deputy Secretary of Defense Robert Work noting that "the Department of Defense seeks a strategy-based, technology oriented approach to maintaining and renewing US military advantage." [10]. Technologically, the third offset focuses on autonomous learning systems, human-machine collaborative decision making, assisted human operations, advanced manned-unmanned system operations, and network-enabled autonomous weapons and high-speed projectiles [11]. In addition to technology, the third offset emphasizes operational and organizational innovation, and innovative military and civilian talent management.

As described above, the goals of this chapter are to analyze the adoption of disruptive technologies and evaluate whether bans and treaties to limit technology proliferation are effective tools. Reviewing the definitions of disruptive technology and examining technology's relationship to revolutions in military affairs and offset strategies provides a context that demonstrate the magnitude and importance of disruptive technology in military affairs. The next section describes the methodology for determining which innovations will be accepted by society. Then, using the framework described below, this chapter will analyze the adoption of various military technologies and map to them bans and/or treaties to limit their use and proliferation.

5 Adapting a Theory from the Life Sciences

Jared Diamond, in his 1997 Pulitzer Prize winning book *Guns, Germs, and Steel*, offers a discussion on the evolution of technology that includes a method for determining which innovations will be accepted by society. This work builds on Diamond's years of research as a professor in geography and evolutionary biology. While his work is concerned with technical innovation adoption as related to the evolution of civilizations, the framework he outlines is also useful for analyzing technical innovation adoption in military affairs. Developing a technology or discovering a new use for a technology is a start. The adoption or acceptance of a novel technology is a critical step to the technology's dissemination and ability to have a disruptive impact. Diamond states that there are at least four factors that influence the technological adoptability: (1) economic advantage (2) social value and prestige (3) compatibility with vested interest (4) observable advantage.

New technologies that are adopted have a relationship to the four factors outlined by Diamond. First, new technology adoption is influenced by the relative economic advantage over existing technology. This characteristic does not have to do with the monetary cost of the technology. It means that a technology must be useful and have a benefit to the user in order to be developed and adopted. Second, technology adoption is influenced by whether the new technology has social value and prestige. A technology that provokes respect or admiration from others is more likely to be adopted. Third, compatibility with vested interests is another factor that determines technology adoption. This third factor bears one of the challenges disruptive technology must overcome—the status quo. A new technology will not be easily accepted if large amounts of effort or money have been poured into the old but proven technologies. Fourth, the ease at which an advantage can be observed is a consideration affecting adoption of a new technology. If the advantage of a new technology is apparent, then it will be more readily adopted. If the advantage takes years to materialize or if the advantage is obscure then the technology may not be accepted. This evaluation of the four criteria must be viewed from the temporal context of when the innovation occurred. The advantages of the crossbow, for example, must be viewed from the context of the Middle Ages, not the Nuclear Age.

Diamond's framework for determining what promotes a technology's acceptance by society is useful because it provides an established structure from which to evaluate dissemination of technology. This chapter will test Diamond's four factor framework against military technologies of the past and present. Additionally, bans/treaties and proliferation will be included as analysis variables. The analysis should reveal any connection between technology adoption, bans and treaties to limit technology use, and technology proliferation.

In order to give the reader background and familiarity with the specific technologies chosen for analysis, the next section of this chapter provides a description of the technologies. The technologies selected for review were chosen because they have the capacity to upset the established way of doing things and the potential to produce unexpected effects—exhibiting the characteristics of disruptive technology. The next section of this chapter will analyze the adoption of various military technologies and map to them (1) bans and/or treaties to limit their use, and (2) proliferation.

5.1 Crossbow

The first technology this chapter will discuss is the crossbow. The crossbow, and its subsequent ban, is one of the earliest examples of a governing body attempting to ban a disruptive technology. Another earlier, but disputed, example is the Greek treaty between city-states Chalcis and Eretria to ban missiles weapons on the battlefield during the Lelantine War around seventh century B.C. [12]. Analyzing the crossbow against Diamond's criteria we find that the crossbow meets all four factors of technology adoption.

First, in the context of the twelfth century, the crossbow provided an economic advantage—it provided benefit to the user because it was a weapon that could be attained and used by those with considerably less financing and less training than a knight. Second, the crossbow provided social value and prestige. Having a crossbow was valuable because it delivered effective means of the handler to engage in offensive and defensive action with comparatively little training: With a crossbow, a peasant could now kill a knight. Third, the crossbow was also able to overcome the status-quo and compatibility with vested interest. The effectiveness over old technologies was so overwhelming that the crossbow was adopted. Finally, the crossbow provided an observable advantage. It was effective. Additionally, the scale of crossbow manufacturing continued expansion and gave early adopters the competitive advantage over other formerly dominant armies [13].

The Second Council of Lateran was convened by Pope Innocent II in 1139 AD. During the time of the Second Council of Lateran, the Catholic Church was one of the most powerful actors in Europe [14]. An important result of the council is Canon 29: The prohibition of use of bows and crossbows against Christians [15]. This prohibition against the use of crossbows is significant because it establishes an early attempt by a governance structure to prohibit use of a game changing technology. The ban failed. The crossbow disseminated and became part of a greater revolution in military affairs (the Infantry Revolution).

Given that some revolutions in military affairs take decades or generations to fully mature, it is entirely likely that the Papal powers of medieval Europe could not fully see the significance of the crossbow technology. The technology's effectiveness and the resulting political and societal implications were profound which is why the weapon was banned. But the crossbow's role in the larger Infantry Revolution was not as visible, partly because it took centuries to complete.

5.2 *Gunpowder*

The crossbow remained the weapon of choice until the introduction of gunpowder [16]. Gunpowder, like the crossbow, first appeared in China in the 9th century [17]. Through many incremental, sustaining improvements, gunpowder proliferated westward to Europe and was recorded as being used in cannons at the Siege of Seville in 1247 [18]. By the mid 1440s, with technological improvements in gunpowder recipes and delivery systems, the Artillery Revolution was underway. Interestingly, in the 16th century, gunpowder also underwrote the Gunpowder Revolution, which took advantage of further improvements of firepower. Gunpowder was a disruptive technology but there were no documented attempts to ban or limit the technology.

Gunpowder meets the criteria outlined by Diamond for technology adoption. First, it provided a benefit to the user. It was a superior technology. Before gunpowder, weapons were designed around the limits of their users' muscular strength. Gunpowder introduced a chemical reaction that delivered results more lethal and at a greater range than swords or bows [19]. Second, it provided social value and prestige. In the early adoption period, only richer parties could afford the new technology [4]. Third, gunpowder was able overcome the status-quo and compatibility with vested interest. Because of gunpowder's effectiveness and range, militaries needed fewer gunman than archers and swordsman [19]. Finally, the gunpowder provided an observable advantage. It was able to overcome the centuries-old dominance of the defense in siege warfare [4].

5.3 *Railroad*

The railroad is one of the technologies that underwrote the Land Warfare Revolution. Railroads were first introduced in England in the seventeenth century and appeared in the United States by 1754 [20]. By the mid-nineteenth century, the utility of the railroad to concentrate forces at great speed and resupply them had been proven [21, 22]. The ability to quickly mobilize and sustain large armies by moving soldiers and supplies at an unprecedented scale and, hence, upset the established way of conducting war. The railroad meets the criteria for being a disruptive technology.

An analysis of railroads shows that it meets the criteria for technology adoption. First, the railroad provided an economic advantage. The commercial and economic sectors of society benefited greatly from the scale and speed of freight transport. Second, the railroad provided social value and prestige. In addition to the commerce facilitated by the railroad and wealth a railroad could generate, having a railroad was an indication that a state was part of the industrial revolution. The investment, materials, labor, and concretization required wealth, sophistication, and belief that undertaking the building a railroad in a given location would be a success. Third, the railroad was also able to overcome the status-quo and compatibility with vested interest. Because of its economic benefit and effectiveness over exiting land transportation technologies, mostly foot, horse, and horse drawn coaches, the railroad was accepted by society. Finally, the railroad provided an observable advantage. Not only was the railroad useful for useful for the transport of goods and people but it also opened new markets and created wealth. The railroad's acceptance by society fits all of the criteria outlined by Diamond.

5.4 *Hollow Point Bullets*

The St. Petersburg Declaration in 1868 prohibited the use of exploding bullets and the Hague Peace Conferences of 1899 and 1907 banned expanding bullets. It is important to note that neither exploding nor expanding bullets were game changing technologies or part of a revolution in military affairs. They can better be described as an incremental advance in technology.

An analysis of the hollow point bullet [23, 24] shows that it meets the four factors for technology adoption. A hollow point bullet was useful because it had a greater range and greater penetration than lead bullets, therefore it did have somewhat of economic advantage. Because it was an incremental advancement in technology and somewhat more effective than lead bullets, the hollow point bullet did carry marginal social value and prestige. The hollow point bullet was also able to overcome the status-quo and compatibility with vested interest. Because hollow point bullets made firearms easier to clean and were generally more accurate, with greater range, they were able to overcome the compatibility with vested interest. Finally, the hollow point bullet provided an observable advantage. This bullet was far more deadly than other bullets and fewer rounds are required to render an enemy hors de combat [23].

5.5 *Chemical and Biological Weapons*

There is a long history of attempts to limit this use of poisons in warfare. The Hague Peace Conferences of 1899 and 1907 attempted to ban poisonous agents and poisoned weapons [25]. The final statements of those international meetings were intended as authoritative declarations but were neither ratified nor implemented by participating states. They did set the stage for future developments in international law and the laws of armed conflict, however. In 1925, the Geneva Protocol sought to further prohibit the use of biological and chemical weapons, after the devastating use of early chemical weapons in World War I. Other significant international legal instruments include the Biological Weapons Convention and the Chemical Weapons

Convention [26], both of which aim to eliminate the use of an entire class of agents in warfare.

The effectiveness of governance strategies prohibiting the use of chemical and biological weapons is debated. Chemical weapons have been deployed in the current Syrian War hundreds of times, by some accounts [27]. The use of chemical weapons by non-state actors is a long-standing threat [28]. Biological weapons have been used domestically in the US by non-states actors and allegedly by states [29, 30]. Neither chemical nor biological weapons alone can be seen as uniquely disruptive, as both have been around for centuries. Both have proliferated far and wide.

An analysis of chemical and biological weapons shows that each meets the criteria for technology adoption. First, chemical and biological weapons provide an economic advantage. They have utility as weapons of war, particularly in reducing capacity of combatants. However, chemical and biological weapons have limited social value and prestige. International norms and protocols have effectively stigmatized them so that only rogue actors are seen as those who would use biological or chemical weapons. Third, because of the longevity and persistence of chemical and biological weapons as tools of warfare (rather than the agents themselves), they effectively have overcome the status-quo and compatibility with vested interest. Finally, biological and chemical weapons can provide an observable advantage. Throughout history, both have been effective weapons for disrupting warfare, for creating casualties among combatants and non-combatants, and engendering fear in a populace.

5.6 Nuclear Weapons

The mid-twentieth century brought game changing technology that immediately changed the character and conduct of warfare. The new technology, nuclear technology, could obliterate a state's economic, political, and social fabric completely. The effectiveness and destructiveness of the nuclear weapons so surpassed all other weapons that nuclear weapons gained utility as deterrent [4]. Nuclear weapons, because of their revolutionary annihilative capability, meet the criteria for providing an economic advantage and providing an observable advantage.

New military doctrine, news systems, and new organization emerged to fully take advantage of the new weapon. Game changing technology underwrote the Nuclear Revolution. Not long after the detonation of nuclear weapons in Japan in 1945 at the end of World War II the international community took steps to limit the technology. Even as progress has been made through multilateral treaties to limit the proliferation and deployment of nuclear weapons, these agreements took decades to develop, and some states are not party to the treaties. Furthermore, the social value and profound prestige of nuclear weapons make them a desirable technology to possess. Nuclear weapons have overcome compatibility with the vested interest in conventional weapons.

The Treaty on the Non-Proliferation of Nuclear Weapons (1968) [31] and the Comprehensive Nuclear-Test-Ban Treaty (1996) [32], in addition to security guarantees, troop deployments, arms sales, nuclear umbrellas, and sanctions threats, have not prevented India, Pakistan, or North Korea from acquiring and testing nuclear weapons. In fact, North Korea tested a nuclear weapon as late as September 2016 [33]. Additionally, Iran continuously edges toward producing nuclear weapons, with or without a nuclear deal [34].

5.7 Landmines

Landmines are a marginally disruptive weapon because they somewhat maintain control of an area, without a person physically being present to assert that control. Landmines were first used on a largescale during WWII. Though widely proliferated, the 1997 Mine Ban Treaty seeks a universal ban on landmines [35].

First, landmines are useful and effective as weapons of war. Second, landmines have limited social value and prestige because international norms and treaties have effectively stigmatized them. Third, as a newer but effective technology, landmines have also overcome compatibility with vested interests because people had not already invested a large amount of money in a similar antecedent. Finally, because of their effectiveness, e.g. destroying a tank, hold territory [36] landmines provide an observable advantage.

5.8 Autonomous Lethal Weapons

Autonomous lethal weapons are remote-control machines operated by civilian or military personnel from a distance. They are potentially disruptive because allow a human to engage in the battlefield from a remote location. Autonomous lethal weapon technology is developing, even as discussions on bans and limits on this weapon system are underway [37].

An analysis of autonomous lethal weapons shows that they meet the criteria for technology adoption. First, autonomous lethal weapons provide an economic advantage because they have utility and potentially spare injury/death of the human handlers. Second, the autonomous lethal weapons provide social value and prestige to those able to obtain and employ the technology. Third, autonomous lethal weapons overcome the status-quo and compatibility with vested interest because they save lives while also being, in some cases, relatively cheap to operate [38]. Finally, autonomous lethal weapons provide an observable advantage: autonomous weapons systems act as a force multiplier, they reach into areas that were previously inaccessible, the efficacy of each warfighter is greater, and they can remove human warfighters from dangerous missions [39, 40].

6 Analysis and Discussion

As the data analysis in Table 2.2 suggests, adoption of military technology adheres to factors outlined for technology adoption by the general population. Every technology analyzed meets all four factors for technology acceptance. Acceptance and adoption is a critical step in technology proliferation. Without acceptance, technology does not disseminate.

The analysis of the cases, as summarized in Table 2.2, also shows that technology, including game changing technology, is unlikely to be completely constrained by treaties. As ineffective as the papal prohibition of the crossbow was eight hundred years ago, this same approach to weapons banning still exists today. The only negligible exception to effectively limiting technology dissemination is the proliferation of nuclear weapons. Even with the intense commitment, resource allocation, monitoring, and intervention, the treaty on non-proliferation of nuclear weapons has not been entirely successful. One surprising point the analysis uncovered is that not all weapons that elicit bans are disruptive. Various technologies, disruptive or not, have elicited bans (Table 2.2).

Creating norms and a governance system that constrains the adoption of game changing technology can be seen as an attempt by those in power to maintain the advantage. However, technology, including game changing technology, does not appear to be successfully constrained by treaties. The crossbow, chemical weapons, and even nuclear weapons serve as examples of how governance has failed to stop the proliferation of a weapons technology or use [41]. As stated by Richard Price:

Table 2.2 Assessment of characteristics of certain technologies with use in warfare and effectiveness of attempts to ban usage

	Disruptive	Relative economic advantage	Social value and prestige	Compatibility with vested interests	Observable advantages	Banned	Proliferation
Crossbow	Yes	Yes	Yes	Yes	Yes	Yes	Yes
Gunpowder	Yes	Yes	Yes	Yes	Yes	No	Yes
Hollow Point Bullets	No	Yes	Marginal/Maybe	Yes	Yes	Yes	Yes
Railroad	Yes	Yes	Yes	Yes	Yes	No	Yes
Chemical Weapons	No	Yes	Marginal/Maybe	Yes	Yes	Yes	Yes
Biological Weapons	No	Yes	Marginal/Maybe	Yes	Yes	Yes	Yes
Nuclear Weapons	Yes	Yes	Yes	Yes	Yes	Yes	Marginal/Maybe
Landmines	Marginal/Maybe	Yes	Marginal/Maybe	Yes	Yes	Yes	Yes
Autonomous lethal weapons	Yes	Yes	Yes	Yes	Yes	Marginal/Maybe	Marginal/Maybe

Yes	Marginal / Maybe	No
(green)	(yellow)	(pink)

> Throughout history, numerous weapons have provoked cries of moral protest upon their introduction as novel technologies of warfare. However…the dominant pattern has been for such moral qualms to disappear over time as these innovations become incorporated into the standard techniques of war. [42]

The pattern of ineffective banning of new technology is important to understanding the process of adopting new technology and to further understanding military affairs. “In order to master an RMA rather than be dragged along by it,” [5] the processes associated with RMA’s must be examined, understood, and improved.

The tension between adoption and prohibition of technology must be considered as part of the balance of power. The utility of treaties may be better viewed as more than only a guarantee against using a weapon. As an editorial in the *Washington Post* suggested, weapons treaties were never an ironclad guarantee that weapons would not be used. Treaties provide a sense of “stability” and “confidence.” [43]. That is, treaties provide stability and are a confidence building measure. With this in mind, effort to find common ground, communicate, and engage diplomatically on terms of agreement on which to cultivate stability should be pushed. The utility of weapon prohibition treaties as balancing should be emphasized, not because prohibition effectively and permanently limits proliferation or use of a technology but because the act of meeting, networking, building relationships, and negotiating provides a forum for interacting and addressing grievances. From this standpoint, governance approaches should be integral to an integrated military strategy for future capabilities development, not the afterthought that attempts to put the genie back in the bottle.

Additionally, to take weapons prohibitions out of the historical prism and into a contemporary context means acknowledging that not all actors are limited by similar adherence to laws and ethical constructs. Take chemical weapons for example. As late as 2017 and 2018, in Syria, President Bahser Al-Assad is thought to have used chemical weapons against his own people [44, 45]. In 2017, North Korea, not a signatory to the Chemical Weapons Convention, used VX nerve agent to commit an assassination of its dictator Kim Jong-un’s half-brother [46]. Moreover, many contemporary threats come from non-state actors. For example, the attacks using chemical weapons (sarin) in Japan in the 1990s were executed by a religious cult [47]. ISIS, a terrorist group is also accused of using chemical weapons in Syria and Iraq [48, 49]. In addition to employing chemical weapons, non-state actors and pariah states also seek nuclear technology and cyber technology for their arsenals.

As strategy should lead technical capability and force development, strategy should similarly lead governance and diplomatic approaches. Using a realist lens, the desire for actors (state or non-state) to possess the game changing technology, despite opportunities and treaties to the contrary, is rational and expected. Weapons prohibition *and* subsequent proliferation must be calculated in every military strategy for future capabilities development. Finally, deterring access of game changing technology to those not part of the existing international state-based governance system is ineffective and requires a separate diligence.

7 Conclusions

This chapter has demonstrated that the pattern of technology prohibitions has weak adherence. The pattern of technology → ban → technology proliferation suggests that there is an element to revolutions in military affairs that begs to be better understood. Better understanding technology adoption and the factors that influence technology dissemination allows for better management of weapons technology proliferation.

This chapter provided a definition of disruptive technology and set disruptive technology within the context of revolutions in military affairs and military offsets strategy. This chapter then analyzed the adoption of various military technologies and mapped to them bans and/or treaties to limit their use and proliferation. Based on the analysis, the final discussion addressed relationships and factors between disruptive technology and governance. Fully exploiting governance tools and making the best, most informed decisions is an imperative aspect in maintaining superiority. Future research should continue to investigate the relationship and interaction between technology and governance, with the goal of enhancing national security strategy.

References

1. Persistent Forecasting of Disruptive Technology (2010) Committee on Forecasting Future Disruptive Technologies, National Research Council. The National Academies Press, Washington DC
2. Bower JL, Christensen CM (1995) Disruptive technologies: catching the wave. Harv Bus Rev 73(1):43–53
3. Christensen CM, Raynor M, McDonald R (2015) What is disruptive innovation? Harv Bus Rev 93(12):44–53
4. Krepinevich AF (1994) Cavalry to computer; the pattern of military revolutions. The National Interest, 1 Sept 1994. https://nationalinterest.org/article/cavalry-to-computer-the-pattern-of-military-revolutions-848
5. Metz S, Kievit J (1995) Strategy and the revolution in military affairs: from theory to policy. Strategic Studies Institute, 27 June 1995. https://ssi.armywarcollege.edu/pdffiles/PUB236.pdf
6. Chapman G (2003) An introduction to the revolution in military affairs. Paper presented at the XV Amaldi conference on problems in global security, Helsinki, Finland, Sept 2003. http://www.lincei.it/rapporti/amaldi/papers/XV-Chapman.pdf
7. Toffler A, Toffler H (1993) War and anti-war: survival at the dawn of the 21st century. Little, Brown and Company, Boston
8. Metz S, Kievit J (1995) Strategy and the revolution in military affairs: from theory to policy. Strategic Studies Institute, 27 June 1995. https://ssi.armywarcollege.edu/pdffiles/PUB236.pdf; Krepinevich AF (1994) Cavalry to computer; the pattern of military revolutions. The National Interest, 1 Sept 1994. https://nationalinterest.org/article/cavalry-to-computer-the-pattern-of-military-revolutions-848; Toffler A, Toffler H (1993) War and anti-war: survival at the dawn of the 21st century. Little, Brown and Company, Boston
9. Hagel C (2014) Defense innovation days. Speech at Southeastern New England Defense Industry Alliance delivered by Secretary of Defense Chuck Hagel, Newport, Rhode Island, 3 Sept 2014. http://archive.defense.gov/Speeches/Speech.aspx?SpeechID=1877

10. Bob Work (2015) Reagan defense forum: the third offset strategy. Speech at Reagan Presidential Library delivered by Deputy Secretary of Defense Bob Work, Simi Valley, CA, 7 Nov 2015. https://www.defense.gov/News/Speeches/Speech-View/Article/628246/reagan-defense-forum-the-third-offset-strategy/
11. Ellman J, Samp L, Coll G (2017) Assessing the third offset strategy. Center for Strategic and International Studies, Mar 2017. https://csis-prod.s3.amazonaws.com/s3fs-public/publication/170302_Ellman_ThirdOffsetStrategySummary_Web.pdf?EXO1GwjFU22_Bkd5A.nx.fJXTKRDKbVR
12. Wheeler EL (1987) Ephorus and the prohibition of missiles. Trans Am Philol Assoc 117: 157–182 (The Johns Hopkins University Press)
13. McNeill WH (1982) The business of war in Europe 1000–1600. In: The pursuit of power. The University of Chicago Press, Chicago, pp 63–81
14. Elliott C (1841) Delineation of Roman Catholicism: Drawn from the authentic and acknowledged standards of the Church of Rome, vol 1. George Lane, New York. https://babel.hathitrust.org/cgi/pt?id=inu.30000112911932;view=1up;seq=9; Ellman J, Samp L, Coll G (2017) Assessing the third offset strategy. Center for Strategic and International Studies, Mar 2017. https://csis-prod.s3.amazonaws.com/s3fs-public/publication/170302_Ellman_ThirdOffsetStrategySummary_Web.pdf?EXO1GwjFU22_Bkd5A.nx.fJXTKRDKbVR
15. Cannons of the Council of Lateran. Papal Encyclopedia Online. http://www.papalencyclicals.net/Councils/ecum10.htm
16. Boothby WH (2009) Weapons and the law of armed conflict. Oxford University Press, New York
17. Andrade T (2016) The gunpowder age: China, military innovation, and the rise of the West in world history. Princeton University Press, Princeton
18. Heard BJ (2008) Handbook of firearms and ballistics: examining and interpreting forensic evidence, 2nd edn. Wiley, West Sussex
19. Curley R (ed) (2012) The Britannica guide to war: war on land. Britannica Educational Publishing, Chicago
20. The Beginnings of American Railroads and Mapping. Library of Congress. https://www.loc.gov/collections/railroad-maps-1828-to-1900/articles-and-essays/history-of-railroads-and-maps/the-beginnings-of-american-railroads-and-mapping/
21. Archer CI (2002) World history of warfare. University of Nebraska Press, Lincoln
22. Van Creveld M (1991) Technology and war: from 2000 BC to the present; a revised and expanded edition. The Free Press: A Division of Macmillan, Inc., New York
23. Berry JF (2010) Hollow point bullets: how history has hijacked their use in combat and why it is time to reexamine the 1899 Hague declaration concerning expanding bullets. Mil Law Rev 206:88–156
24. Hamilton JB (1898) The evolution of the dum-dum bullet. Br Med J 1(1950):1250–1251
25. Schindler D, Toman J (eds) (1988) The laws of armed conflicts, Third Revised and Completed Edition, 4th edn. M Nihjoff, Dordrecht, pp 53–56
26. 1925 Protocol for the Prohibition of the Use in War of Asphyxiating, Poisonous or Other Gases, and of Bacteriological Methods of Warfare. (The Geneva Protocol), United Nations Office for Disarmament Affairs. http://disarmament.un.org/treaties/t/1925. See also "On the Prohibition of the Development, Production and Stockpiling of Bacteriological (Biological) and Toxin Weapons and on Their Destruction" (The Biological Weapons Convention), United Nations Office for Disarmament Affiars. http://disarmament.un.org/treaties/t/bwc and "Convention on the Prohibition of the Development, Production, Stockpiling and Use of Chemical Weapons and on their Destruction," (The Chemical Weapons Convention). http://disarmament.un.org/treaties/t/cwc
27. Schneider T, Lutkefend T (2019) Nowhere to hide: the logic of chemical weapons use in Syria. Global Public Policy Institute, Feb 2019. https://www.gppi.net/media/GPPi_Schneider_Luetkefend_2019_Nowhere_to_Hide_Web.pdf
28. Kosal ME (2009) Near term threats of chemical weapons terrorism. In: Russell J, Wirtz J (eds) Globalization and WMD proliferation: terrorism, transnational networks and international security. Routledge, London, pp 63–78

29. Tucker J (ed) (2000) Toxic terror: assessing terrorist use of chemical and biological weapons. MIT Press, Cambridge
30. Frischknecht F (2003) The history of biological warfare. Sci Soc 4(Suppl 1):S47–S52
31. 1968 Treaty on the Non-Proliferation of Nuclear Weapons (NPT). United Nations Office for Disarmament Affairs. https://www.un.org/disarmament/wmd/nuclear/npt/
32. 1996 Comprehensive Nuclear Test-Ban Treaty. United Nations Office for Disarmament Affairs. https://www.un.org/disarmament/wmd/nuclear/ctbt/
33. Hunt K, Kwon KJ, Hanna J (2016) North Korea claims successful test of nuclear warhead. CNN, 10 Sept 2016. http://www.cnn.com/2016/09/08/asia/north-korea-seismic-activity/
34. The Historic Deal that Will Prevent Iran from Acquiring a Nuclear Weapon. The White House archive, President Barack Obama. https://obamawhitehouse.archives.gov/issues/foreign-policy/iran-deal
35. Convention on the Prohibition of the Use, Stockpiling, Production and Transfer of Anti-Personnel Mines and on their Destruction, United Nations, 18 Sept 1997. https://www.un.org/Depts/mine/UNDocs/ban_trty.htm
36. Croll M (1998) The history of landmines. Pen and Sword, London
37. UN: Key Action on 'Killer Robots' International Move toward Possible Ban, Human Rights Watch, 16 Dec 2016. https://www.hrw.org/news/2016/12/16/un-key-action-killer-robots
38. Steinbuch Y (2017) War against ISIS now involves shuttlecock grenades dropped by drones. New York Post, 16 Mar 2017. http://nypost.com/2017/03/16/war-against-isis-now-involves-shuttlecock-grenades-dropped-by-drones/
39. Etzioni A, Etzioni O (2017) Pros and cons of autonomous weapons systems. Mil Rev 72–18 (2017). https://www.armyupress.army.mil/Portals/7/military-review/Archives/English/pros-and-cons-of-autonomous-weapons-systems.pdf
40. Unmanned Systems Roadmap 2007–2032. Department of Defense, 10 Dec 2007. https://fas.org/irp/program/collect/usroadmap2007.pdf
41. Garcia D, Herz M (2016) Preventative action in world politics. Glob Policy 7(3):370–379
42. Price R (1995) A genealogy of the chemical weapons taboo. Int Organ 49(1):73–103 (Winter 1995)
43. Weapons treaties look ready to unravel, and that's a big problem. Washington Post, 30 Nov 2016. https://www.washingtonpost.com/opinions/weapons-treaties-look-ready-to-unravel-and-thats-a-big-problem/2016/11/30/2ffcf11e-b5ad-11e6-b8df-600bd9d38a02_story.html
44. Third Report of the Organization for the Prohibition of chemical Weapons-United Nations Joint Investigative Mechanism. United Nations Security Council, 24 Aug 2016, http://www.un.org/ga/search/view_doc.asp?symbol=s/2016/738
45. Death by Chemicals: The Syrian Government's Widespread and Systematic Use of Chemical Weapons. Human Rights Watch, 1 May 2017. https://www.hrw.org/report/2017/05/01/death-chemicals/syrian-governments-widespread-and-systematic-use-chemical-weapons
46. Berlinger J (2017) VX nerve agent used to kill Kim Jong Nam, police say. CNN, 27 Feb 2017. http://www.cnn.com/2017/02/23/asia/kim-jong-nam-vx-nerve-agent
47. Cronin AK (2003) Terrorist motivations for chemical and biological weapons use: placing the threat in context. Congressional Research Service, Report RL31831, 28 Mar 2003. https://fas.org/irp/crs/RL31831.pdf
48. Schmitt E (2016) ISIS used chemical arms at least 53 times in Syria and Iraq, report says. New York Times, 21 Nov 2016. https://www.nytimes.com/2016/11/21/world/middleeast/isis-chemical-weapons-syria-iraq-mosul.html?_r=0
49. Wille B (2017) ISIS accused of unleashing chemical weapons in Mosul: Iraqi government and humanitarians should do more to protect civilians. Human Rights Watch, 6 Mar 2017. https://www.hrw.org/news/2017/03/06/isis-accused-unleashing-chemical-weapons-mosul

Emerging Life Sciences: New Challenges to Strategic Stability

Margaret E. Kosal

Abstract This chapter explores the potential for new biotechnologically-enabled weapons to compete with nuclear weapons as far as effect on strategic stability, and assesses whether the assumptions in traditional strategic stability models are still valid when applied to such scenarios, and how changing capabilities and adversaries may shape approaches to arms control, verification, and monitoring. When thinking about biotechnology from a security perspective, anticipating the types of security threats that may emerge as science and technology advance, the potential consequences of those threats, the probability that adversaries will obtain or pursue them, adversarial intent, and potential effect on strategic stability is necessary. The range and spectrum of possible capabilities and actors are expanding. The most recent addition to the genome-editing arsenal is CRISPR, a bacteria-derived system that is among the simplest genome-editing tools. The CRISPR-Cas9 system—and emerging variants on the system—enables unprecedented control and ease when editing the genome. With parallels to remote 'command and control' of the genome, this is one aspect that makes the technology different from earlier gene-editing methods. Contemporary analyses of emerging technologies often expose tenuous links or disconnections between technical realities and mainstream scholarship. How, when, where, and in what form the shifting nature of technological progress may bring enhanced or entirely new capabilities, many of which are no longer the exclusive domain of a single nation-state, is contested and requires better analytical tools to enable assessment and inform policy choices. This work is hardly the only one to consider the biosecurity implications of CRISPR, gene-editing, and broader issues of biotechnology. As far as is known, it is the only one to address these emerging life sciences technologies in the context of nuclear strategic stability and implications for balance of power, arms control, and international security.

M. E. Kosal (✉)
Georgia Institute of Technology, Atlanta, GA, USA
e-mail: margaret.kosal@inta.gatech.edu

M. E. Kosal (ed.), *Disruptive and Game Changing Technologies in Modern Warfare*, Advanced Sciences and Technologies for Security Applications,
https://doi.org/10.1007/978-3-030-28342-1_3

1 Introduction and Scoping

In thinking about the potential security implications of emerging technologies, one of the first areas to highlight is the changing strategic environment. We are no longer in a purely post-Cold War international security environment, but we are in an environment that combines those characteristics with greater uncertainty in the nature of our adversaries and in their capabilities. As the lone superpower remaining even in the face of rising China, the United States has an unprecedented level of military and other technological capabilities. Because of these asymmetric advantages, particularly in traditional military power, adversaries may seek nontraditional and unconventional means to affect the United States and our allies.

In the post-Cold War environment, the most technologically advanced military power no longer guarantees national security. Globalization and the information revolution, including the Internet and other communication leaps—have led to much greater visibility into the availability and potential for technology [1]. New technological developments have become accessible and relatively inexpensive to a larger number of nations and are within the grasp of non-state actors: advanced technology is no longer the domain of the few [2].

In the 21st century, both nation-states and non-state actors may have access to new and potentially devastating dual-use technology [3, 4]. A risk that has been highlighted by prominent public figures, policy-makers, and other voices [5–9].[1] In a scholarly context, this work is hardly the only to have thought about the biosecurity implications of CRISPR, gene-editing, and broader issues of biotechnology.[2] As far as is known, it is the only one to address substantively these emerging life sciences technologies in the context of nuclear strategic stability and implications for balance of power, arms control, and international security.

Emerging innovations within today's most cutting-edge science and technology (S&T) areas are cited as carrying the potential of bringing about both the near term capabilities, as well as those that might appear scientific fictions, closer. Those S&T areas include biotechnology, including systems biology and new gene editing techniques like CRISPR-cas9; nanotechnology, including meta-materials for unprecedented stealth capabilities; robotics, including artificial intelligence and autonomy; the cognitive neurosciences; new materials and processes, including additive manufacturing (aka 3-D printing); and the intersection of each with the cyber world of information and communications technologies (ICTs).

Communication of those new discoveries is occurring faster than ever, meaning that the unique ownership of a piece of new technology is no longer a sufficient position to maintain a competitive edge. The information revolution and globalization themselves have been major drivers. It is widely regarded that recognition of

[1] "New scientific and technological developments relevant to the Convention: Some examples," BWC Preparatory Committee, Eighth Review Conference of the States Parties to the Convention on the Prohibition of the Development, Production and Stockpiling of Bacteriological (Biological) and Toxin Weapons and on Their Destruction, BWC/CONF.VIII/PC/WP.18, August 5, 2016.

[2] For example, see [10–14].

the potential applications of a technology and a sense of purpose in exploiting it are far more important than simply having access to it today. Technological surprise has and will continue to take many forms. A plethora of new technologies that are under development for peaceful means but may have unintended security consequences and will certainly require innovative countermeasures. For example, tremendous developments in biotechnology have occurred since the advent of recombinant DNA and tissue culture-based processes in the 1970s. If the potential for biotechnology to affect fundamental security and warfighting doctrines had been more clearly recognized twenty years ago, the situation today could be very different. The threat of biological weapons' use—from both states and non-state actors—currently presents a threat that is difficult to predict and for which traditional solutions are increasingly less effective. Defense against such threats requires strategic foresight to be valuable.

For this chapter, I chose to approach the issue through three questions, starting with the fundamental query: *What technological scenario would compete with nuclear as far as effect on strategic stability?* If there is not a scenario that satisfies that inquiry, then the think-piece would move to consideration of permutations on strategic stability or a null set discussing why those technologies are not relevant. The next question: *Are the technological assumptions in traditional strategic stability models still valid?* This section considers how the assumptions for nuclear-based stability might differ and the role of uncertainty in this regard. The last question I will consider: *How do technological changes shape approaches to verification and monitoring, and will emerging technologies present new compliance challenges?* This section explores how the material differences associated with emerging biological agents will affect the political processes of creating, implementing and executing an international regime, including implementation and verification. What would an organization that is responsible for such verification look like in terms of orientation, composition, and responsibility? This last question also relates most closely to the challenges of modern arms control.

2 What Technological Scenario Could Compete with Nuclear Weapons as Far as Effect on Strategic Stability?

From the domain of the life sciences, biology, and biotechnology, traditional biological weapons do not fulfill such a scenario, nor would the re-emergence of small pox. Humanity survived with smallpox until 1980; we could do it again. It would be painful and disruptive, but it would not likely impact strategic stability. This would not be the kind of biological weapons that dominated 20th century offensive programs.

Yes, infectious disease can impact domestic, regional, and perhaps global stability [14–17], e.g., outbreak of Ebola in West Africa in 2014 [18–20]. The extent to which that epidemic was due to or exacerbated by structural and capacity inadequacies of the three nations most effected (Liberia, Sierra Leone, and Guinea) because of civil wars and other long-standing political and economic factors has been well-documented [21]. Cases were reported, including human to human infections, in Nigeria, Mali, Senegal, and the US. The latter states were able to isolate/quarantine those affected and respond adequately to halt the spread of the disease.

In order to satisfy a scenario that altered the strategic balance, something would have to threaten in a way that was comparable to or exceeded the potential destruction of nuclear weapons use. What would that look like? For this chapter, I suggest that change to the human germline, or heritable human genetic modifications, may satisfy that scenario. More specifically, two scenarios emerge as possible; first, germline changes that cause mass sterility of a nation or a biotechnologically-engineered weapon that causes fast, non-treatable death, such as by over-stimulation of the nervous system, could potentially compete with nuclear weapons in this regard.

From a pre-CRISPR-era perspective, a biotechnological advancement with the hypothetical capability to cause catastrophic death has been demonstrated in what has come to be known as the Australian Mousepox Experiment. In that work published in 2011, a gene known to suppress the immune system was spliced into a contagious virus creating a strain so powerful that it killed even those mice inoculated against the virus. A team of scientists initially inserted Interleukin-4 (IL-4), a protein involved in immune response, and mouse egg protein into the mouse pox virus with the goal of triggering an immune response against mouse eggs to make the animals infertile [22]. As hypothesized, the animals became infertile; however, the modified virus also suppressed normal anti-viral cell-mediated cytokine responses—meaning that the animals' livers were destroyed and approximately half of the mice died, even though they had previously been vaccinated against mousepox. Previously, IL-4 had been observed to cause increased pathogenicity in some viruses (e.g., influenza), but not in others.

The work was subsequently repeated by a different, independent team of scientists as part of an effort to develop a counter-measure. In this iteration, all of the mice died [23]. The team reportedly observed similar results with cowpox, which can infect humans, and rabbitpox viruses [24]. The latter work has been presented at scientific conferences, but it has not been published in open, research literature.

To satisfy a scenario that challenges nuclear weapons for strategic stability, the agent would need to be highly transmissible; even more so than small pox. In epidemiology, a useful measure of the severity of a contagious disease outbreak is the basic reproduction number, R nought (R0), which is defined as the average number of secondary cases (i.e., additional individuals infected) of an infectious disease arising from a typical case in a group of people [25]. This number is dependent upon transmissibility, or the probability an infected individual transmits the infection to someone else, the duration of infectiousness, or how long an infected person remains "contagious," and how likely infected individuals will interact with uninfected individuals, or roughly population density. The measure is a critical factor in mathematical models that attempt to describe the spread of biological agents during a natural disease outbreak or bioterrorist incident, and should be used for public health and other planning and educational simulations. The R0 for smallpox (the vaccinia variola major virus) is 1.8–3.2. By comparison, R0 for untreated pneumonic plague caused by Yersinia pestis is 0.8–3.0; for polio the R0 is 5–7, and for Ebola-causing filovirus R0 is 1.34–1.8 [26]. The R0 for the virus that causes measles is generally found to be 12–18 [25], although in some situations, such as unvaccinated populations, it has been found to be greater than 40 [27]. Effective vaccination can reduce R0 for all diseases to zero. The mortality for untreated measles is significantly lower than for diseases such as smallpox, Ebola, or untreated pneumonia, however.

Thinking about limiting the threat of biological agents is more complicated than any single value. For this chapter, invoking the concept of basic reproduction number, R0, is used to suggest that a target that might challenge nuclear weapons in strategic stability, as in the first scenario, is not likely to be one that is expected. It is likely to be a chimeric engineered species, such as one that leverages the contagiousness of measles.

The mousepox IL-4 experiments illustrate conceptually how mass sterility might be hypothesized from an experimental design perspective, though they are not technically exact "recipes" for extrapolation to human systems. The potential for creativity in this respect shouldn't be dismissed, but also should not be over-hyped. As currently performed, the experiments also require the researcher to individually inject and dose test subjects, which means deployable concepts of operation for distribution are not part of the experimental procedure.

The mouse pox experiments were performed using traditional laborious techniques. More recently, three techniques have been discovered and are being explored for clinical applications. One such technique, CRISPR, has received the most attention, but another system—Zinc Finger Nucleases (ZFN)—has progressed further in clinical applications, including treatment of sickle-cell disease [28], antitrypsin deficiency, Parkinson's and others [29, 30].

CRISPR is the most recent addition to the genome-editing arsenal and is a bacteria-derived system that uses RNA molecules that recognize specific human DNA sequences.[3] CRISPR stands for clustered regularly interspaced short palindromic repeats (CRISPR) and is associated with specific proteins, such as Cas9, an endonuclease (enzyme) that uses a guide-sequence within the RNA duplex. The RNAs act as guides, matching the nuclease to corresponding locations in the human genome. CRISPR-Cas9 is among the simplest genome-editing tool to use because it relies on RNA-DNA base-pairing rather than designing and synthesizing particular proteins that bind specific DNA sequences.

Already at least two technically simpler techniques than CRISPR-Cas9 have been reported: a family of smaller Cas enzymes, SaCas9, that overcomes some challenges with delivering Cas9 in vivo, given they are smaller [32]; and Cpf1, a smaller version of Cas9 derived from a different bacteria, a dual-nuclease enzyme that can perform both the pre-RNA processing and DNA cleavage activities [33, 34].

Successful germline editing has been performed in animal models such as fish [35], rats [36], cattle [37], sheep [38], and pigs [39]. In one example, germline editing was used to modify embryonic development of the neural crest in amphibians and

[3]For an excellent overview of the CRISPR-Cas9 system by two of the discovers, see [31].

these modifications were observed in approximately 22% of descendants [40]. The neural crest (NC) is a unique cell population that are pluripotent, i.e., can give rise to several different cell types and highly migratory. The neural crest cells form a variety of cell types, including the peripheral nervous system and facial skeletons.

Unsuccessful modification of human germline was reported in 2015 [41]. A group of Chinese scientists reported using CRISPR-Cas9 gene-editing techniques to modify human embryos. Their paper, published in the Chinese journal *Protein & Cell*, came as little surprise to the scientific community, but it did renew debate about what types of gene-editing research should be performed and how to potentially limit this research. The paper describes how Guangzhou-based scientists used the CRISPR-Cas9 system to cut DNA in human embryos and then attempted to repair it by introducing new DNA [42]. The team used non-viable embryos obtained from fertility clinics, in which the eggs had been modified so that these experiments would not result in a live birth. The publication also raises questions about the appropriate way to disseminate such experimental work.

In November 2018, Chinese scientist He Jiankui announced that twin baby girls had been born after he used CRISPR to edit the DNA of two human embryos, which were successfully carried to term after re-implantation in the mother's body [43]. This work garnered attention because the gene-editing that He Jiankui and his team performed affected the germline; that is, the modification can be carried on to children that the two girls, named Lulu and Nana, might have in the future. Less well known is the fact that He Jiankui reported that another Chinese woman is pregnant with a CRISPR gene-edited embryo—a fact which has been tacitly confirmed [44].

As of this writing, the experimental procedure and research claimed by He has not been verified, nor has it been subjected to intensive scientific review (separate from legal and ethical review). Based on what He has publicly released, it appears that the editing process was not as precise as initially described; i.e., "off-target" mutations were observed among other technical criticisms [45]. While CRISPR has been lauded for its specificity (compared to other techniques), off-target or "promiscuous" behavior by CRISPR-Cas9 has been reported [46–48].

Following the controversy around He's announcement, Chinese President Xi Jinping called for new regulations on gene editing: "Under the proposed measures, released on Tuesday, technology involving gene editing, gene transfer and gene regulation would be categorized as "high-risk" and placed under the authority of the State Council, China's cabinet." [49]. Such categorization would be equivalent to or potentially even more restrictive than Select Agent Rules in the United States.

Subsequent research by scientists unassociated with the Chinese researchers found that the mutation that He and his team claim they performed correlate with increased mortality [50]. Earlier work has also shown that the edited gene has a role in memory and cognition [51, 52], which has led some to hypothesize that it might lead to human enhancements [53].

Exemplifying scientific research that is not unique to any single nation, CRISPR technology is also of interest to a Russian scientist who announced that he was pursuing similar gene-editing experiments and intended to use CRISPR for germline editing of humans in June 2019 [54]. Russia has recently thumbed its metaphorical nose at international norms with the use of unscheduled, military-grade nerve agents against former government affiliates (Sergei Skripal and his daughter, Yulia Skripal) living in the UK. If Russia is pursuing broader efforts to erode the post-WWII liberal international order, of which the Skripal case arguably is exemplar, the Russian scientists' announcement would be an indication of such. Alternatively, this could be an individual scientist sending up a test balloon to see how Putin responds. Regardless, the demonstrated interest and intent to use CRISPR gene-editing techniques is ongoing and not limited to any single nation-state.

3 Are the Technological Assumptions in Traditional Strategic Stability Models Still Valid?

For Cold Warriors, strategic stability was a military rationale. Schelling and Wohlstetter-esque "stability of mutual deterrence" evokes strong connotations of stable and unstable equilibrium from physics. Strategic stability was all about surviving a first nuclear attack and then credibly being able to respond with a massive retaliatory nuclear strike. The Cold War paradigm sought strategic stability through parity of nuclear arsenals in terms of capabilities, numbers, and conceptual permissiveness of limited nuclear war fighting and conformity of intent.

While some scholars have asserted that biological weapons themselves are more suitably used as a strategic deterrent than nuclear weapons because "effective deterrence requires only a small possibility of great destruction [55]," that view remains a minority perspective.

Assumptions regarding technological prowess and sources, especially of development and use, have changed. At the technical level, CRISPR is still in the development-side of applied research in many ways. Control of exactly how many cells are modified can be difficult. Increasing the dose of nuclease used would increase the likelihood that the mutated gene will be corrected, but it also raises the risk of gene cuts being made elsewhere in the genome.

Operationally, the concepts of operations for deployment of a gene editing system are limited to clinical settings. The method for introducing gene therapy is essentially the same as is used for Assisted Reproductive Technology (ART) through in vitro fertilization-embryo transfer (IVF-ET), which is not a procedure adaptable to large-scale use in a non-permissive or conflict setting.

Changes to the germ-line that enable new capabilities are largely—but not completely—speculative, and how they would be accomplished is as well. Are the potential capabilities of CRISPR-based or like systems similar to the threshold from atomic to hydrogen-based weapons? Or is it more akin to transition from conventional explosives in mortars to nuclear weapons? Or does neither analogy apply?

Like nuclear weapons biological weapons are indiscriminate weapons. In a population lacking a countermeasure, infectious transmissible biological weapons may be more indiscriminate. When thinking about biological agents, a critical variable in considering their use and effectiveness is uncertainty. The uncertainties are potentially much larger with biological agents than with the deployment of nuclear weapons. Because of the large range of uncertainties, developing strategic postures and policies to defend against their use is more fraught with risk. In international security, states will hedge to worst-case scenarios, especially in the face of a potential existential or strategic threat.

Other issues that should be further examined include assumptions regarding vulnerability and the US' ability to retaliate. The 2010 US Nuclear Posture Review (NPR) noted that previously the United States "reserved the right to employ nuclear weapons to deter CBW attack on the United States and its allies and partners." [56]. However, this NPR included a change from a nuclear response to "any state eligible for the assurance that uses chemical or biological weapons against the United States or its allies and partners would face the prospect of a devastating conventional military response—and that any individuals responsible for the attack, whether national leaders or military commanders, would be held fully accountable." This change is largely due to confidence in the deterrent threat of conventional military force.

The threat of advanced biological weapons was highlighted in detail in the 2010 NPR as being one potentially foreseeable exception to the new deterrence policy: "Given the catastrophic potential of biological weapons and the rapid pace of biotechnology development, the United States reserves the right to make any adjustment in the assurance that may be warranted by the evolution and proliferation of the biological weapons threat and U.S. capacities to counter that threat." [56]. The most recent NPR, released in 2018 as an Executive Summary, did not include any such language regarding response to non-nuclear attack or advanced biological weapons [57].

As far as the impact on military structure, the recent shift of proponency and lead for the Counter-Weapons of Mass Destruction (C-WMD) mission from US Strategic Command (STRATCOM) to US Special Operations Command (SOCOM) signals the ongoing efforts by the defense establishment to respond to traditional WMD threats. Although threats from WMD are neither new nor unrecognized at the highest levels of the U.S. Government (USG) and Department of Defense (DOD), remarkable gaps and inconsistencies between strategic-level policy and operational capabilities persist [58]. While a prevention strategy is laudable and important, the disparity between strategy and the required operational capabilities and capacities necessary for securing, interdicting, and eliminating WMD reveals potential gaps that must be recognized and accounted for to ensure a credible deterrent posture. Future threats, especially biological, are likely to be more complicated than current or past conceptions.

These are some of the questions that need to be explored more fully. Others exist as well, such as questions surrounding assumptions with respect to survivability and for whom such scenarios as highlighted above convey offensive advantage. What can be done domestically to increase the deterrent power of the US with respect to these potential weapons? Is there enormous advantage to the side that first uses/deploys such technology and, if so, how does the US reduce it? What actions are complementary to pre-emptive use of this technology, and, therefore, contrary to the stability of mutual deterrence? And which actions are not complementary when it comes to advanced biotechnology? The answers to these questions are unlikely to come from a single individual or even a small group of people. Diversity of experience and background are critical to answering these questions.

4 Verification and Monitoring

One doesn't need to re-articulate (or regurgitate) the litany of challenges, to put it diplomatically, associated with biological verification regimes [59–63]. While nuclear weapons have their own challenges, they can be counted and nuclear weapons proliferation can be detected and verified more easily as compared with verification in the biological (and chemical) realms.

Advanced biotechnologies already present new compliance challenges. When thinking about biotechnology from a security governance perspective, anticipating the types of security threats that may emerge as science and technology advance, the potential consequences of those threats, the probability that adversaries will obtain or pursue them, adversarial intent, and potential effect on strategic stability is essential. Many times, it's less the specific technologies and rather surrounding geopolitical and economic uncertainties that effect their use.

Operationally (from a policy perspective), the domestic Select Agent Rules are predicated on what you start with, i.e., it's a specific list of viruses, bacteria, rickettsia, and toxins that have special requirements to work with them. The proposed Gain of Function (GOF) policies are based on what one ends up with, i.e., what unspecified microbiological agents can be made to do (above and beyond what they can do courtesy of evolution). CRISP-cas9 is one technique that allows one to go from one to the other. As discussed above, Zinc-finger nucleases (ZFN) are another example of such technologies. As such, focusing upon one technique alone is not useful in considering governance of this broad area. More than one set of rules is needed to govern this space because the object of the regulations is different from the starting material, the actual tools used, and the potential resulting product.

The majority of states internationally do not have explicit legislation in place permitting or forbidding genetic engineering in humans. Most consider such research experimental and not therapeutic. However, in nations with policies regarding inheritable genetic modification, it has been prohibited by law or by other regulatory mechanisms. This consensus is most visible in Western Europe, where 15 of 22 nations prohibit the modification of the germ line [64, 65]. Although the US has not officially prohibited germline modification, the US National Institutes of Health's Recombinant DNA Advisory Committee explicitly states that it "will not at present entertain proposals for germ line alterations, but will consider proposals involving somatic cell transfer." [66, 67].

The CRISPR-Cas9 system—and emerging variants on the system—enable unprecedented control and ease when editing the genome. It's somewhat analogous to remote 'command and control' of the genome and is what makes the tools novel and different from earlier gene-editing methods. Potential impacts of this tool on biomedicine and human health are vast, including beneficial applications for enabling gene and cell replacement therapies, drug discovery through functional genomic screens, and simplification of the production of disease models that permit validated therapy targets and increased efficiency in testing drug efficacy. In practical terms, the potential applications of CRISPR-Cas9 and the challenges and pitfalls associated with the technology are still to be determined; particularly the pitfalls of the technology with implications for international security. Governance that addresses such uncertainties, while not hindering research, is tough.

The broader biosecurity and nonproliferation communities, along with Congressionally-charted Committee findings [68], have recognized that in the 21st Century, biological weapons are sometimes (but not always) cheaper, easier to produce, more widely available, within the capabilities of an increasingly large number of people with access to minimal technical skills and equipment and more concealable dual-use technologies. The potential synergies between biotechnology and other emerging technologies, such as nanotechnology, big data analytics, and the cognitive neurosciences, not only suggest tremendous potential promise for advancement in technology for consumers and defense applications, but also raise new concerns [69]. A driving concern is that in the 21st century, both nation-states and non-state actors may have access to new and potentially devastating dual-use technology. Advanced technology is no longer the domain of a few. Biological weapons are perceived as relatively cheap and accessible, especially when compared to the obstacles in attaining and developing nuclear weapons.

Reducing the risk from state-based misuse of biotechnology for biological proliferation will mean consideration of the highly transnational nature of biotechnology research and development. Traditional and innovative new approaches to nonproliferation and counterproliferation are important policy elements to reduce the risk of malfeasant application of technology. Robust international agreements lower the risk of terrorist applications by eliminating legal routes for states and terrorists to obtain agents, precursors, or weaponization materials, and by minimizing possible transfers from state to non-state actors through theft, deception, or other means. Efforts to strengthen the international regime to control transfers of dual-use materials and equipment are also important. The highly transnational nature of biotechnology research and development is a major consideration in reducing the risk of misuse of biotechnology for weapons.

Creating, implementing, and executing effective verification entails two critical challenges, in this scholar's view, and they are not necessarily the usual ones cited. The two main challenges are scope and structure. How the first, scope of verification, is to be resolved will directly affect the challenges of the second.

Is the scope of BWC verification constrained to mid-twentieth century technology, biological weapons, and weaponization techniques? Are we only looking for that handful of traditional microbiological species that are CDC Category A, B, & C agents? Should verification efforts be limited to the traditional pathogenic bacterial and viral-based weapons? The risks associated with weaponized *Bacillus anthracis*, etc. have not gone away. Rather, in terms of likelihood of biological weapons proliferation, they remain arguably the largest risks and the mid-twentieth century capabilities have arguably diffused the furthest.

Some would look to the BWC text and assert that a verification regime unequivocally could not be limited in any such way. At the same time, the most intrusive verification regime agreed upon by the international community is based on a delineation of inspection targets. The Chemical Weapons Convention (CWC) has its 'Schedules of Chemicals' as a critical part of its inspection and verification regime. That argument can only go so far, however. From an International Relations realist perspective, the stronger counter-argument against articulation of CWC-like schedules of biological agents, materials, and equipment is that it would effectively provide a list of what to avoid if one wanted to cheat to potential proliferators.

Globalization and the information revolution have made new technological developments accessible and relatively inexpensive to many nations and within the grasp of non-state actors. Advances in biotechnology and information technology have been driven by needs for improved biomedical products, public health, or industrial applications. The Internet and other communication leaps have led to much greater visibility into the availability and potential for technology. Either intentionally or not, these advances have also fostered the proliferation of knowledge, as well as spurred interest in the creation of novel non-traditional offensive uses of advanced technology. Without a Scientific Advisory Board (SAB) or institutionalized means to provide advice on truly emerging technologies with implications for biological proliferation, a BWC verification protocol would have difficulty avoiding becoming an artifact of, at best, late 20th Century microbiology and engineering.

The most effective institution to achieve verification for 21st Century challenges of biological weapons proliferation is not likely to resemble the dominant—and admittedly often quite effective—institutions of Cold War era nuclear verification. That is most certainly a model from which the biological weapons nonproliferation community has and should continue to learn; but, this verification institution should not be one on which the international community relies too heavily or on which any nation rests its confidence for securing any emerging technology, biological or otherwise. The perils—owned as this commentator's characterization—of modeling approaches to limiting biological weapons proliferation too closely to those used to limit nuclear weapons have been eloquently discussed previously [70]. Too often, policymakers have given into the temptation (or been pushed) to incorporate approaches to biological agents under the auspices of strategies associated with nuclear weapons, particularly with respect to material control. Yet this undermines and simplifies many distinct and differing characteristics of biological agents and nuclear material. One such method often attempted in this regard is a focus upon pathogen security, or securing and denying access to the materials necessary in developing biological weapons. With nuclear weapons, securing and safeguarding fissile material is a reasonable goal because of key characteristics of the ingredients, notably the distinct uses of highly enriched uranium, the origin of the substance, and the technologies associated with the processes.

Those same characteristics that make nuclear weapons amenable to tracking are what make biological weapons material difficult to monitor and verify. These characteristics include presence in nature, production costs, diversity of material, and other legitimate uses. While many active members of the biosecurity and biological nonproliferation communities recognize this, one does not have to travel far to encounter someone espousing a biological agents variant of 'Materials, Protection, Control, and Accountability' programs. The structure should learn from but not be wedded to nuclear models.

An effective verification regime for reducing the threat of biological weapons in the 21st century is more likely to resemble an intelligence activity than a traditional arms control office. The structure of an institution intended to take on the potential full scope of biotechnology must be able to absorb and respond to uncertainties to be most effective, particularly if it is to have a deterrent effect with respect to potential proliferators. While the BWC, and other international mechanisms, does help create international "norms" that condemn the use of biological agents for state or non-state use as a weapon, it does not constitute a deterrent by itself. A verification regime could provide the global support for deterrent strategies if credible attribution were achieved. For such a regime to be credible, the uncertainties would need to be acknowledged and methods to address them formulated, tested, and institutionalized. While intellectually provocative and useful in strategic planning contexts, delineating lists of likely future biological weapons threats is a risky game at best. The reasonable and acceptable uncertainties of biological verification should be explored, and done so in more than a cursory way. The challenges and the real work are ultimately in the details of resolving the scope and structure.

Rather than the typical focus on securing biological materials and laboratories from misuse, other recommendations and programs may be more effective and include: bio-surveillance and early detection capabilities along with global laboratory and research cooperation—a 21st century extension of the Nunn-Lugar Cooperative Threat Reduction programs.

The issue of a state (or in the lesser probability, a terrorist group) utilizing advanced biological weapons against another state is a mounting concern due to the fact that dangerous pathogens have been known to cause deadly effects, yet little weapons deterrence research addresses methods of dealing with the threat of biological weapons and even less so with deterring bioterrorism. Bridging the gap between the life sciences and social sciences is a crucial element in devising implementable and executable strategies that can lead to successful deterrence of bioweapons. Similarly, thinking about verification and international arms control regimes can be explored as part of new approaches to strategic deterrence in the 21st Century.

5 Conclusions and Recommendations

The dual use conundrum applies to all modern technologies, but has become a greater concern due to other characteristics of the changing strategic environment. Reducing the risk from misuse of technology will mean consideration of the highly transnational nature of the critical technology required. Traditional and innovative new approaches to nonproliferation and counter proliferation are important policy elements in reducing the risk of malfeasant applications of technology that may enable advanced weapons or increase the availability of the production or dissemination of biochemical agents to a much wider group of actors. Efforts to strengthen existing international regimes to control transfers of dual-use materials are also important [71]. Verification still remains a technical as well as diplomatic challenge and the role of international agreements and cooperative programs in the 21st Century is a contested intellectual and policy field.

In his inaugural address, in which the formation of a North Atlantic Security Pact (NATO) was first mentioned, US President Harry S. Truman asserted as one of the last of the four major foreign policy goals of his administration that "we must embark on a bold new program for making the benefits of our scientific advances and industrial progress available for the improvement and growth of underdeveloped areas." [72]. In addition to being part of the plan for rebuilding Europe, science and technology were intrinsic to domestic economic development and defense goals. Scientific exchanges remain a core area of NATO and can be seen in many other defense-oriented agreements, such as bilateral, trilateral, and multi-lateral Memoranda of Understanding (MOU), partnerships, and other arrangements.

The instruments of science diplomacy include tools such as MOUs and other official government-to-government interactions: the classic tools of traditional Track I diplomacy. Science diplomacy has perhaps made the biggest impact in foreign policy as a part of Track II diplomatic efforts: informal diplomacy between individuals who are not officially empowered to act on behalf of the state but are acting in accordance with a state's foreign policy goals and interact through dialogue, exchanges, cooperative programs, or other means as part of increasing cooperation and transparency or in decreasing conflict among states. Track II efforts with nuclear physicists and other scientists during the Cold War are legendary.

In many ways, nuclear diplomacy of the Cold War may be argued as the pinnacle of Track II science diplomacy. Overall, Track II science diplomacy has been an underutilized tool since then, which may be ironic considering that, since the early 1990s, the world has become increasingly technologically-dependent and technology has enabled the spread, at an unprecedented rate, of scientific knowledge, capabilities, and materials globally.

Initiated following the end of the Cold War, a core component of Cooperative Threat Reduction (CTR) efforts were focused upon redirecting the offensive or weapons-based knowledge and skill-sets of scientists in former Soviet states to defensive or peaceful aims, including Track II science diplomacy. CTR has traditionally, and by statute of associated public funding, focused upon reducing the risks from nuclear, biological, and chemical weapons. One can envision a role for science diplomacy beyond the former Soviet states and beyond those weapons as part of a pro-active, 21st Century Cooperative Threat Reduction program; for example, one might imagine a program in partnership with Russia to engage Pakistani and Indian scientists and engineers in cooperative threat reduction with respect to the misuse of bio-nanotechnology or synthetic biology. As a model policy leveraging science diplomacy to increase global security, CTR offers opportunities in the diplomatic realm, in the engagement of scientists and engineers, and for study by international affairs scholars.

In the 21st Century major barriers to effective science diplomacy include three major risks: not being relevant, not being strategic, and not being at the table. Science is increasingly complicated and complex. The ability to translate and make relevant the role and importance of science to foreign policy aims is critical. While there are notable exceptions, often this goal is not best accomplished by active research scientists. It is also not often accomplished well by traditional Foreign Service Officers. In the global information age, there is a critical need for champions of science diplomacy and for a cohort of individuals who are capable of bridging the divide across technical and foreign policy arenas. In the US, the former is institutionalized and purportedly embodied in the Science and Technology Advisor to the Secretary of State (STAS).

Technical experts are vital, and lack of expertise can set back efforts by years. The ability to bridge those gaps and work between the technical and the political realms is sometimes over-looked. While having a highly-regarded, chaired "Professor X" helicopter in from his lab at "Top Ten University" provides credibility and insight into cutting edge work, and sometimes provides notoriety, the critical work of synthesizing, analyzing, and determining the implications for national and international security rarely emerges. Once the metaphorical spotlight has been used to illuminate some issue, the science diplomats and other inside and outside the government who possess some mix of technical and policy expertise are responsible for creating, implementing, executing, and assessing the results. There are few, if any, examples in which a single agency, never mind a single individual, possess the solutions at the global scale. It requires empowered and resourced teams of individuals, and increasingly those teams are multi-national, i.e., requiring those with international experience, understanding, and backgrounds.

Being invited to the metaphorical—and often literal—table where decisions on foreign policy are made (or at least getting a seat in the second row along the wall) is the third critical piece to effective science diplomacy. This requires individuals who can bridge the realms of science and policy. It also requires institutional means by which such individuals can gain access. One example, albeit not the only, is the American Association for the Advancement of Science (AAAS) Science and Technology (S&T) Policy Fellows Program, which enables individuals to spend one or two years placed in the State Department as temporary employees. It is the means by which doctoral level scientists are literally able to get their feet in the door of the State Department and beyond the visitor's center unescorted. Being familiar conveys opportunity, understanding, advantage, credibility, and trust.

While technically-trained individuals have made their way to foreign policy, that single US program has been the most effective in developing and sustaining the transition of high-performing scientists and engineers into foreign policy. The forty-five year old program has enabled over 2,000 individuals with doctoral level degrees in scientific or engineering fields to spend a year or two working within the Executive or Legislative Branch of the US Government. Among the more than a dozen federal agencies in which AAAS S&T Policy Fellows may spend their time, the State Department and US Agency for International Development (USAID) have two of the strongest and most established programs. There are also vibrant S&T Fellows within the Departments of Defense and Homeland Security. They also have become an effective and valued means by which doctoral level scientists become practitioners of Track I science diplomacy as permanent US State Department employees and throughout the US federal government and Congress. This is one example of an institution that plays a critical role at the intersection of science, technology, and international affairs.

Like other technological breakthroughs of the late twentieth and early twenty-first century, such as artificial intelligence, the ubiquitous extent of 'cyber' through information and communications technologies, nanotechnology, and the cognitive sciences, much of the concern regarding the potential offensive applications of gene-editing is highly speculative and based on worst-case scenarios. The technical and operational veracity of scenarios varies highly from the robust to Hollywood-like fantasy. In the post-Amerithrax world, particularly of the industrialized global north, worst-case scenarios garner easy media attention and can inadvertently drive policy decisions. Choices can be made today, and policy can be implemented in the near future that are likely to shift the balance in favor of maximizing the beneficial and minimizing the negative effects on global security.

Included in those choices are flexible approaches to nonproliferation and counter-proliferation, which are important policy elements in reducing the potential risks posed by emerging biotechnologies. Past methods for other technologies that don't take into account the international nature of the science and technology industry are not adequate. Any international regime must be interdisciplinary in focus, cognizant of the multi-polar post-Cold War world, and appreciate the role of private funders, commercial development, and transnational corporations. To be clear, there is much to learn from and leverage in existing arms control and nonproliferation institutions. These starting points and history are valuable and US leader should learn from them; they are not necessarily predictive, however. Regardless, the challenges in this arena are primarily political rather than technical.

In the last decade or so, scientists – especially in the western world – have been compelled to step up and take on more and more responsibility for security in large part because of inadequacy and lack of governance solutions, which are largely a product of the 20th Century Cold War, based on norms that predate that, as well as the inability of political measures to adequately deal with the changing threat environment. The ability to navigate and affect policy at the interface of science and international affairs is a way in which we could have immense value in addressing the threats posed by biotechnological weapons and their proliferation.

References

1. Rennstich KJ (2008) The making of a digital world: the evolution of technological change and how it shaped our world. Palgrave MacMillan, New York
2. Unclassified Key Judgments of the National Intelligence Estimate (2007) Prospects for Iraq's Stability: a Challenging Road Ahead. Office of the Director of National Intelligence, https://www.dni.gov/files/documents/Newsroom/Press%20Releases/2007%20Press%20Releases/20070202_release.pdf
3. Biotechnology Research in an Age of Terrorism. National Academies Press, Washington, D.C. (2004)
4. Globalization, Biosecurity, and the Future of the Life Sciences. National Academies Press, Washington, D.C. (2006)
5. Bill Gates warns tens of millions could be killed by bio-terrorism. The Guardian (UK), 18 Feb 2017. https://www.theguardian.com/technology/2017/feb/18/bill-gates-warns-tens-of-millions-could-be-killed-by-bio-terrorism
6. Clapper JR (2016) Worldwide Threat Assessment of the US Intelligence Community, Statement for the Record to the Senate Armed Services Committee, 9 Feb 2016. https://www.dni.gov/files/documents/SASC_Unclassified_2016_ATA_SFR_FINAL.pdf
7. Letter to the President, President's Council of Advisors on Science and Technology, Executive Office of the President, Nov 2016. https://obamawhitehouse.archives.gov/sites/default/files/microsites/ostp/PCAST/pcast_biodefense_letter_report_final.pdf
8. Reeves RG et al (2018) Agricultural research, or a new bioweapon system? Science 362(6410):35–37
9. Dando M (2016) Find the time to discuss new bioweapons. Nature 535:9. https://www.nature.com/news/find-the-time-to-discuss-new-bioweapons-1.20206
10. Vogel KM, Ouagrham-Gormley SB (2018) Anticipating emerging biotechnology threats: a case study of CRISPR. Polit Life Sci 37(2):203–219
11. Gronvall G (2018) The security implications of synthetic biology. Survival 60(4):165–180
12. Oye KA et al (2014) Regulating gene drives. Science 345(6197):626–628
13. Koblentz GD (2004) Pathogens as weapons: the international security implications of biological warfare. Int Secur 29(1):84–122
14. Peterson S (2002) Epidemic disease and national security. Secur Stud 12(2):43–81
15. Elbe S (2002) HIV/AIDS and the changing landscape of war in Africa. Int Secur 27(2):159–177
16. McInnes C, Lee Kelley (2006) Health, security and foreign policy. Rev Int Stud 32:5–23
17. Letendre K, Fincher CL, Thornhill R (2010) Does infectious disease cause global variation in the frequency of intrastate armed conflict and civil war? Biol Rev 85(3):669–683
18. Waddington C (2014) Ebola outbreak in Guinea: a different type of regional stability threat. Africa Confl Monthly Monit 2014(5):47–51
19. Heymann DL et al (2015) Global health security: the wider lessons from the west African Ebola virus disease epidemic. Lancet 385(9980):1805–1806
20. Omoleke SA, Mohammed I, Saidu Y (2016) Ebola viral disease in West Africa: a threat to global health, economy and political stability. J Public Health Africa 7(1):27–40
21. Kalra S et al (2014) The emergence of Ebola as a global health security threat: from 'Lessons Learned' to coordinated multilateral containment efforts. J Glob Infect Dis 6(4): 164–177
22. Jackson RJ et al (2001) Expression of mouse interleukin-4 by a recombinant ectromelia virus suppresses cytolytic lymphocyte responses and overcomes genetic resistance to mousepox. J Virol 75:1205–1210
23. Chen N et al (2011) Poxvirus interleukin-4 expression overcomes inherent resistance and vaccine-induced immunity: pathogenesis, prophylaxis, and antiviral therapy. Virology 409(2):328–337
24. MacKenzie D (2003) US develops lethal new viruses. New Sci. https://www.newscientist.com/article/dn4318-us-develops-lethal-new-viruses/
25. Guerra FM et al (2017) The basic reproduction number (R0) of measles: a systematic review. Lancet 17(12):e420–e428

26. Fine PE (1993) Herd immunity: history, theory, practice. Epid Rev 15:265–302
27. Wallinga J, Lévy-Bruhl D, Gay NJ, Wachmann CH (2001) Estimation of measles reproduction ratios and prospects for elimination of measles by vaccination in some western European countries. Epid Infect 127:281–295
28. Porteus M (2006) Mammalian gene targeting with designed zinc finger nucleases. Mol Ther 13:438–446
29. Mansilla-Soto J, Riviere I, Boulad F et al (2016) Cell and gene therapy for the beta-thalassemias: advances and prospects. Hum Gene Ther 27(4): 295–304
30. Jo Yi, Kim H, Ramakrishna S (2015) Recent developments and clinical studies utilizing engineered zinc finger nuclease technology. Cell Mol Life Sci 72(20):3819–3830
31. Doudna JA, Charpentier E (2014) The new frontier of genome engineering with CRISPR-Cas9. Science 346(6213):1077–1088
32. Ran FA et al (2015) In vivo genome editing using Staphylococcus aureus Cas9. Nature 520:186–191
33. Zetsche B et al (2015) Cpf1 is a single RNA-guided endonuclease of a class 2 CRISPR-cas system. Cell 163(3):759–771
34. Fonfara I et al (2016) The CRISPR-associated DNA-cleaving enzyme Cpf1 also processes precursor CRISPR RNA. Nature 532:517–521
35. McKenna A et al (2016) Whole organism lineage tracing by combinatorial and cumulative genome editing. Science 353(6298):aaf7907-11
36. Yoshimi K, Kaneko T, Voigt B, Mashimo T (2014) Allele-specific genome editing and correction of disease-associated phenotypes in rats using the CRISPR-cas platform. Nat Commun 5(4240):1–9
37. Heo TY et al (2015) CRISPR/Cas9 nuclease-mediated gene knock-in in bovine-induced pluripotent cells. Stem Cells Dev 24:393–402
38. Han H et al (2014) One-step generation of myostatin gene knockout sheep via the CRISPR/Cas9 system. Front Agric Sci Eng 1: 2–15
39. Kang Q et al (2014) Improving pig genetic resistance and muscle production through molecular biology. In: Proceeding of 10th world congress genetics applied livestock production. https://asas.org/docs/default-source/wcgalp-proceedings-oral/362_paper_10607_manuscript_1526_0.pdf?sfvrsn=2
40. Efficient genome editing of genes involved in neural crest development using the CRISPR/Cas9 system in *Xenopus embryos*. Cell Biosci 6(22) (2016). http://cellandbioscience.biomedcentral.com/articles/10.1186/s13578-016-0088-4
41. Liang P et al (2015) CRISPR/Cas9-mediated gene editing in human tripronuclear zygotes. Protein Cell 6(5):363–372
42. Cyranoski D, Reardon S (2015) Embryo editing sparks epic debate. Nature 520:593–595
43. Normile D (2018) Shock greets claim of CRISPR-edited babies. Science 362(6418):978–979
44. He J (2019) China condemns 'baby gene editing' scientist. BBC News, 21 Jan 2019. https://www.bbc.com/news/world-asia-46943593
45. Wang H, Yang H (2019) Gene-edited babies: what went wrong and what could go wrong. PLOS Biol. https://journals.plos.org/plosbiology/article?id=10.1371/journal.pbio.3000224
46. O'Green H, Abigail S, Sega DJ (2015) How specific is CRISPR/Cas9 really?. Curr Opin Chem Biol 29:72–78
47. Tsai SQ et al (2015) GUIDE-seq enables genome-wide profiling of off-target cleavage by CRISPR-Cas nucleases. Nat Biotechnol 33:187–197
48. Chapman JE, Gillum D, Kiani S (2017) Approaches to reduce CRISPR off-target effects for safer genome editing. Appl Biosaf J ABSA Int 22(1):7–13
49. Cyranoski D (2019) China set to introduce gene-editing regulation following CRISPR-baby furore. Nature. https://www.nature.com/articles/d41586-019-01580-1
50. Wei X, Nielsen R (2019) CCR52-Δ32 is deleterious in the homozygous state in humans. Nat Med 25:909–910
51. Zhou M et al (2016) CCR5 is a suppressor for cortical plasticity and hippocampal learning and memory. eLife 20(5):e20985

52. Joy MT et al (2019) CCR54 is a therapeutic target for recovery after stroke and traumatic brain injury. Cell 176(5):1143–1157
53. Regalado A (2019) China's CRISPR twins might have had their brains inadvertently enhanced. MIT Technol Rev. https://www.technologyreview.com/s/612997/the-crispr-twins-had-their-brains-altered/
54. Cyranoski D (2019) Russian biologist plans more CRISPR-edited babies. Nature. https://www.nature.com/articles/d41586-019-01770-x
55. Martin S (2002) The role of biological weapons in international politics: the real military revolution. J Strateg Stud 25(1):76
56. Nuclear Posture Review Report (2010) viii, United States Department of Defense. https://dod.defense.gov/Portals/1/features/defenseReviews/NPR/2010_Nuclear_Posture_Review_Report.pdf
57. Nuclear Posture Review Executive Summary (2018) United States Department of Defense. https://media.defense.gov/2018/Feb/02/2001872877/-1/-1/1/EXECUTIVE-SUMMARY.PDF
58. Kosal ME (2018) WMD strategy gap: capacities, capabilities, and collaboration. PRISM 7(3):50–67
59. Smithson Amy E (1999) Tall order: crafting a meaningful verification protocol for the biological weapons convention. Polit Life Sci 18(1):79–85
60. Littlewood J (2005) The biological weapons convention: a failed revolution. Ashgate Publishing, Aldershot
61. Littlewood J (2012) The biological weapons convention. In: Faure G (ed) Unfinished business: why international negotiations fail. University of Georgia Press, Athens, pp 107–129
62. Sims NA (2006) Toward the BWC review conference: diplomacy still in the doldrums. Disarmament Diplomacy, vol 82. http://www.acronym.org.uk/old/dd/dd82/82ns.htm
63. Hart J, Trapp R (2012) Science, technology, and the biological weapons convention. Arms Control Today 42(8):15–21
64. Araki M, Ishii T (2014) International regulatory landscape and integration of corrective genome editing into in vitro fertilization. Reprod Biol Endocrinol 12(108). http://www.rbej.com/content/12/1/108
65. Tebas P et al (2014) Gene editing of CCR68 in autologous CD4 T cells of persons infected with HIV. N Engl J Med 370:901–910
66. Flotte TR (2015) Therapeutic germ line alteration: has CRISPR/Cas9 technology forced the question? Hum Gene Ther 26(5):245–246
67. NIH-RAC. NIH Guidelines for research involving recombinant or synthetic nucleic acid molecules (NIH Guidelines), Apr 2016. http://osp.od.nih.gov/office-biotechnology-activities/biosafety/nih-guidelines
68. World at Risk: The Report of the Commission on the Prevention of WMD Proliferation and Terrorism, Dec 2008. https://www.absa.org/leg/WorldAtRisk.pdf
69. Kosal ME (2009) Nanotechnology for chemical and biological defense. Springer Academic Publishers, New York
70. Ostfield M (2009) Pathogen security: the illusion of security in foreign policy and biodefense. Int J Risk Assess Manag 12:204–221
71. Kosal ME (2008) U.S. policies to reduce the threat of chemical terrorism, in 9/11 + 6 Initiative Foreign Policy Priorities for a Secure America, The Partnership for a Secure America, May 2008. http://www.psaonline.org/downloads/CHEMICAL%20report%208-28-08.pdf
72. Truman HS (1949) Inaugural address. Online by Gerhard Peters and John T. Woolley, The American Presidency Project, 20 Jan 1949. http://www.presidency.ucsb.edu/ws/?pid=13282

Additive Manufacturing (AM) and WMD Proliferation

Andrew Conant

Abstract Additive manufacturing (AM) or 3D printing technology is a rising industry with applications that traverse all sectors of the economy. A variety of users can use 3D printing commercially or recreationally to make objects in plastic and metal, thus it has caused concern among the nuclear proliferation community in its dual-use capability to help acquisition by non-weapons states or non-state actors. Despite the concern, current AM capabilities give little cause for alarm. What AM possesses in flexibility, it lacks in depth; AM has limitations in size, material strength, and cost of objects compared to traditional manufacturing methods. Spread of sensitive nuclear technologies is not possible with the technology in the near future. Delivery systems are more worrisome yet their actualization probability remains low. The United States and international community should work together to continually examine AM capabilities in the near term and begin to update export control mechanisms, re-examine signatures of proliferation for the intelligence community, and promote collaborative efforts between the AM technical community and the public sector to alert of disruptive ability of the technology.

1 Introduction

> "Additive manufacturing poses challenges for militaries and governments with the de facto responsibility to keep war-making tools out of the wrong hands." John Mattox, National Defense University Center for the Study of Weapons of Mass Destruction [1].

Experts have touted the onset of the fourth industrial revolution [2, 3], marked by technologies that integrates the digital age (third industrial revolution, following steam power and electrification) into society and even the human body. Technologies in the fourth industrial revolution include: artificial intelligence, nanotechnology, advanced robotics, the Internet of Things, and advanced manufacturing capabilities, especially additive manufacturing. In the post-digital age, unprecedented

A. Conant (✉)
Georgia Institute of Technology, Atlanta, GA, USA
e-mail: aconant3@gatech.edu

M. E. Kosal (ed.), *Disruptive and Game Changing Technologies in Modern Warfare*, Advanced Sciences and Technologies for Security Applications,
https://doi.org/10.1007/978-3-030-28342-1_4

manufacturing techniques could alter the current manufacturing paradigm and supply chains [4, 5]. Traditionally, engineers have designed and created products according to subtractive manufacturing techniques, i.e., removing material from a fixed-size object. Economies worldwide have perfected these techniques to optimize the speed and cost of the production of goods.

Recent improvements in additive manufacturing, i.e., adding layers to create objects, have arisen in the past several decades. The private sector has capitalized on its use in creating quick prototypes of products, which has given rise to a function-based synonym for 3D printing, rapid prototyping. A 3D printing machine will add layer-by-layer material of some plastic, resin, or metal. Common methods to produce these objects include extrusion (unwinding a wire-shaped feed material), stereolithography (shining light on surface to bond molecules of a liquid polymer together), laser sintering or melting (focusing a laser on metal powder to bond molecules and successively adding powder layers on top). These methods require a computer-aided design (CAD) file as an input; a computer program or the printer itself will deconstruct the image into many cross-sectional layers to be used as steps for the printer.

What are the current capabilities of 3D printers? For commercial 3D printers, they spread the gamut of sizes and prices. The cost ranges from several hundred to a few thousand dollars, and the feed filament costs approximately twenty to thirty dollars per kilogram [6]. Most household 3D printers are relatively small, and their application is only relevant to relatively small objects less than one half of a meter in one dimension [7]. The physical limitations render it useful only for low-quality objects, such as gears, screws, household tools, etc. However, even "household" 3D printers can have resolution up to the sub-millimeter scale. A plethora of websites contains ready-to-print stereolithography (STL) files [8–10], which feed into most 3D printers or allow conversion to a similar format.

Industrial 3D printers, as expected, come with higher costs yet more robust capabilities. The majority owners of higher-tech 3D printers include Department of Energy national laboratories, defense contractors, and large companies such as General Electric and Hewlett Packard. Oak Ridge National Laboratory (ORNL) printed the first-ever 3D printed car, a 2014 Shelby Cobra with their Big Area Additive Manufacturing (BAAM) machine [11]. Lockheed Martin uses additive manufacturing to produce prototypes and parts for satellites and fighter jets; it also operates several AM innovation centers and an AM machine that have printed objects of various sizes for spacecraft components [12]. Raytheon, another defense contractor, successfully printed the components and assembled a small missile [13].

The end use for many commercial and industrial applications include rapid prototyping of objects and making objects that are traditionally difficult to manufacture. Should it be timely and cost-effective, it has the potential to replace stapled manufacturing processes such as casting, molding, and forming. Because each layer is added successively with AM as opposed to relying on the hardening or shaping of feed material, orientations that are traditionally challenging to manufacture become either achievable, more efficient, or both. Current 3D printing technology lacks time efficiency on a large scale, therefore the technology is most applicable to rapid prototyping. The Shelby Cobra took six weeks to go from the start of printing to drivable car [14]; most weapons and single-use systems will have a higher threshold for performance. The effort to produce a single sample object requires less effort in machining. Furthermore, even if an actor or organization does not possess the technology, 3D print shops and services, although not ubiquitous, are available.

2 Relevance and Background

Additive manufacturing has several implications for US national security. First, the 3D printing technology is of dual-use in nature. It can be utilized benevolently to make products such as prosthetics, implants, and car parts, but it can also be used to make potentially harmful objects. For example, an organization called Defense Distributed circulated a design file for a handgun called the Liberator [15]. The US State Department asked the organization to recant the file, which prompted Defense Distributed to sue stating its violation of several constitutional amendments. The US Government won the case due to the focus of its argument on national security [16]. Governments may have difficulty with sensitive objects such as the Liberator because it is challenging to regulate its spread under the International Traffic in Arms Regulations, which aims to limit the proliferation of traditional arms as their proliferation could enable terrorism and proliferation of weapons of mass destruction (WMD). Counterterrorism and counter-WMD proliferation remain two of the most vital strategic objectives to protect the US homeland, allies abroad, and overall global stability.

Because 3D printers maintain the flexibility to print objects of virtually any shape, this new technology requires exploration in its ability to aid in the proliferation of WMD. The threat of a rogue state or non-state actor obtaining WMD relies on their ability to secure sensitive chemical, biological, radiological, or nuclear material (CBRN) and to obtain the necessary components.

It is hypothesized that AM could disrupt traditional acquisition means of the materials needed to create a WMD. Rather than purchasing the required technology, an actor could print the pieces themselves. An actor must gain knowledge to produce the pieces, but the knowledge to produce pieces via 3D printing is lower than that using traditional manufacturing methods. Design, pre/post processing techniques, and process surveillance are not as labor- or knowledge-intensive with AM, although not to understate the importance of tacit knowledge. Lockheed Martin and other corporations have also demonstrated that techniques such as laser sintering and melting allow production of higher-strength metals [17]. The facile procurement of computer files over the Internet permits almost any actor to have access to these files. The files are not so easily detectable and the end use of the eventual 3D-printed object can be unclear. Evidently, weapons of mass destruction pose a threat to US citizens at home and abroad as well as threaten the security offered by the strategic position of the United States [18]. An easier acquisition of these weapons decreases the significance of the US deterrent threat.

In addition to the relative ease in fabricating machined parts, widespread use of AM could make it difficult for the counter-WMD strategy of the United States. It decreases the size of facilities that could be used to create WMD, thereby "rendering detection by international inspectors or national intelligence agencies much harder." [19]. AM is touted as a technology that can bring 3D printing to each household, therefore it is not unfathomable to assume that AM weapons production could be dispersed throughout a wider area or in multiple, smaller buildings. This phenomenon could increase the security dilemma for the United States; the probability of successful detection of a covert WMD program decreases and the transparency of weapons manufacturing decreases.

Current thinking on the evolution of additive manufacturing also raise two potential long-term impacts on US security interests: energy efficiency and economic dominance. Increases in energy efficiency maintain positive economic and environmental impacts on the United States; citizens save money and negative environmental impacts are reduced. Additive manufacturing, as compared to subtractive manufacturing, produces little waste due to the nature of the technology. Subtractive methods can use as little as 5% of the input material whereas the additive methods can use 98% or more of the input material in its final product; additive methods have also been shown to use approximately 50% less energy to produce parts [20]. If these statistics are true, the United States has a lot to gain from this technology. Another potential consequence of international implementation of the technology is that it could reduce the dominance of the US manufacturing sector. The US relies on protecting its infrastructure to maintain economic security in international markets [21]. 3D printers could decrease the infrastructure threshold, equalizing the capabilities among states. Both of these claims are of little significance currently as AM has not grown to the scale of traditional manufacturing and thus will not be examined here. Little evidence proves that these are immediate concerns, but the actualization of these speculations could impact long-term US national security.

3 Methodological Approach

This chapter attempts to analyze the potential effects of projected AM advancements on WMD proliferation. This analysis categorizes sensitive CBRN and supplemental technologies necessary to utilize a WMD. The focus of the analysis is the feasibility of fabricating each piece of technology with additive manufacturing methods under assumption of current capabilities and future advancements.

What does this chapter not attempt to answer? The long-term economic implications of additive manufacturing and potential changes to international manufacturing sections is highly speculative and dependent on the evolution of the technology in the coming decades. Discussion of the security of traditional firearms is also not discussed here. Proliferation of traditional munitions, except for those that can serve as systems of a WMD, is not relevant here. The literature has exhausted security implications of the increased availability of 3D-printed traditional weapons [22], and these weapons have little potential to be disruptive in military encounters. Of little importance to this analysis is the role of AM technology to support criminal activities, such as drug smuggling or intellectual property theft; such instances are of more relevance to law enforcement than international military conflict. Disruptions to the traditional military balance of power in this discussion will turn to WMD acquisition.

What are the key areas of weapons of mass destruction in which additive manufacturing could facilitate or hinder acquisition? In which WMD areas do the threats lie? This chapter classifies WMD threats into CBRN and their supplemental systems, including production and delivery technologies. A state or non-state actor to acquire WMD assumes a change in the regional or international balance of power, which would curtail the power and threaten interests of the United States and its allies.

This methodology requires a detailed analysis of the technologies described above and their capabilities. Prospective capabilities are assumed to be linked to motivation, i.e., WMD-seeking actors will be attracted capabilities that permit easier access to a WMD. This metric will be analyzed with respect to questions such as: Are current weapons technologies able to be printed with current AM technology? Are there size or material restrictions? Is it any more transferrable than current methods? Are there limits on their spread from international agreements? This methodology will be qualitative but also will include some semi-quantitative aspects. Feasibility of printing objects equates comparable mechanical properties of materials, including strength, temperature limits, porosity, corrosiveness, etc. Intangible metrics such as knowledge, both theoretical and tacit, will also be factored into the analysis.

Independent variables include capabilities of current and near-term improvements in additive manufacturing capabilities. The dependent variable will be the effect on WMD acquisition by an actor seeking those weapons or on the ability of a WMD-garnering actor to prevent the WMD acquisition of the WMD-seeking actor. The former dependent variable has a higher impact on the disruptive nature of the technology. WMD proliferation is assumed to be more disruptive to the balance of power, particularly within regions.

4 Novelty

Little exploration of this technology and its impact on WMD and counter-WMD has been performed. A prominent work detailing the threat of additive manufacturing to the spread of nuclear weapons specifically is an article by Kroenig and Volpe in the October 2015 issue of *The Washington Quarterly* [19]. The authors argue that 3D printing enables WMD-proliferation because it requires little technical knowledge and potential facilities that could produce WMD-sensitive parts can be widespread and impossible to detect. Although there are logical conclusions, he simplifies the technology without further examining it and how it would be realistically implemented by a WMD-seeking actor along with the international regimes that could re-analyze proliferation threats with respect to AM technology. They simply assume that rogue states or non-state actors will covertly pursue the technology. The authors do not answer the question of how, i.e., what would a covert AM-driven nuclear WMD program look like?

Another gap in the existing literature is more speculative and draws on comparison to successfully disruptive technologies such as the Internet and personal computers [23]. In both instances, technologies gave informational and entrepreneurial power to the individual. Experts have created analogies between these technologies and additive manufacturing, but they fail to dive past the surface level. They believe that the individual nature of a these technologies warranted its success, and therefore additive manufacturing will follow a similar trajectory to that of personal computers. They assume advancements in AM are inevitable and exponential, hence disruptive over a short period. Many articles cite the attention and investment AM has received over recent years, with AM innovation centers surfacing in the United States, Europe, and Asia, as the main indicator of its potential [24]. Some scholars, however, have projected that AM rests at the top of its hype curve and that it requires great technological and institutional demands to overtake traditional manufacturing methods [25]. Some assessments state that "the ability to produce weapons outside traditional fabrication channels also carries additional challenges" yet fail to dig deep into the feasibility and investment necessary to actualize that path [26].

Other sources have focused on the application of additive manufacturing in the military industrial complex [1] and the spread of 3D-printed traditional munitions [27]. The former does have implications in the speed of the military to actualize a product, while the latter does pose real international security concerns. Both struggle to connect these changes to their potential impacts on weapons of mass destruction. The former article states that there are "catastrophic consequences [with] the prospect of additive manufacturing technologies being applied to produce weapons of mass destruction." Generalizations are made about how quantities are lower for successful production and the facilities are easier to hide. There lacks an understanding of the detailed implementation should a state or actor pursue a WMD through these means and which technologies are most sensitive should an actor pursue an AM capability. What facilities should military forces seek and target? How can the international community limit these capabilities through export control? What are indicators of proliferation through this technology?

Current research fails to adequately acknowledge or discredit the role of additive manufacturing as it relates to WMD acquisition by rogue states and non-state actors. Although concessions exist that the technology is not up to par to be viewed as immediately threatening, scholars tend to shortcut to the end point where AM is the ideal disruptive technology due to ideal characteristics that it has yet to currently achieve. A technical breakdown of the technologies is necessary to examine the practical use of the technology to analyze the true threat to US national security interests.

5 Nuclear and Radiological Applications

The nuclear proliferation threat relies on two main components of the nuclear fuel cycle: enrichment and reprocessing capabilities. Because highly-enriched uranium can only be produced with enrichment technology and weapons-usable plutonium can only be produced with reprocessing capabilities, these are the primary technologies of concern for WMD proliferation. Of these two sensitive stages of the nuclear fuel cycle, one must be implemented for the successful acquisition of a nuclear bomb. The exception to that is the case where a fabricated nuclear bomb is stolen, however this risk is not heightened with the advent of advanced manufacturing technologies.

Enrichment capabilities are used to increase the fissile content of natural or low-enriched uranium to weapons-grade uranium. The most current case of uranium enrichment for WMD-seeking purposes is Iran. As part of the Joint Comprehensive Plan of Action (JCPOA), Iran had agreed to decrease its number of centrifuges. They agreed to downsize from 19,000 to 6,100 centrifuges, and all of the remaining gas centrifuges are to be of the IR-1 type. Models of this centrifuge range from IR-1 to IR-8, all specified to similar dimensions. They all have similar orders in terms of heights and diameters, no more than 0.65 m and no more than 2.5 m, respectively [28]. With size constraints, these centrifuges could theoretically be 3D printed with a moderately large 3D printer. Components that require the smallest resolution in a gas centrifuge, e.g., two millimeters, such as the molecular pumps and motor stators, could also be made [29].

A major problem with centrifuges is that they require highly corrosive-resistant materials. Uranium hexafluoride, the form of the uranium in the centrifuge, is highly corrosive to most metals. The Nuclear Suppliers Group, a collection of countries that support export controls on nuclear-sensitive technologies, suggests that maraging steel or strong aluminum alloys be used for rotating components to avoid corrosion; neither of these materials are used extensively outside sophisticated laboratories. Variations of maraging steel and aluminum alloys have been commercially and experimentally listed as below the NSG threshold for ultimate tensile strength necessary for a gas centrifuge component [30]. Even if the strength of the material met NSG standards, exporting it to a non-weapons state would disregard international treaties. If a NSG country wanted to disregard the agreement, it could do so without any consideration or use of AM technology.

In addition to these technical limitations, logistical limitations also exist. The theoretical time required to additively manufacture, assemble, and arrange hundreds or thousands of centrifuges would render it impractical. AM has solely been proven effective, disregarding economics, for small-scale production or prototyping. An actor deciding to pursue these weapons would more likely decide to invest in a "tried and true" method, such as through the experience of the AQ Khan network [31]. Furthermore, an enrichment facility requires the cascading, or joining, of hundreds or thousands of centrifuges to increase their utility. Such a facility could likely be detected through surveillance methods, as was the case with the Natanz facility in

Iran [32]. A compelling case would be if a new centrifuge configuration could be designed to fit in a smaller space, yet this novelty would not be due to improvements in additive manufacturing.

Reprocessing capabilities, on the other hand, were developed to chemically separate uranium from plutonium in spent nuclear fuel. Reprocessing technology has been the preferred route for several proliferating countries, including the ostensibly-proliferated countries of India, Pakistan, and Israel. The main ingredient in nuclear reprocessing is already-used nuclear fuel. Many processes exist to separate plutonium, but the most widely used is the Purex (plutonium uranium extraction) process. Purex is a solvent extraction method that uses nitric acid to separate plutonium and uranium by their oxidation states [33]. Albeit a straightforward chemical process, Purex implementation requires expertise in nuclear-related disciplines.

Nuclear fuel to be reprocessed will be at high levels of radioactivity, therefore advanced hot cells are a necessary technology. Criticality safety experts are needs to ensure subcritical, and therefore safe, results of the process. Radiation shielding materials, such as concrete, are also required to limit dose to workers at the facility. These materials and expertise are the main barriers to constructing a reprocessing plant with enough throughput to fabricate a plutonium weapon. Slabs of concrete and the complicated, large components for hot cells needed to handle nuclear fuel are not feasible hurdles for AM to surmount. Traditional manufacturing methods have the advantage in this regard; AM would not be worth the financial and knowledge investment to develop a reprocessing facility. This excludes the assumption that an actor has access to a significant quantity of fissile material and therefore must bypass the current nonproliferation regime, which has been deemed as "successful in retarding the global spread of nuclear weapons." [19].

The production of a radiological weapon shares challenges with that of crude nuclear weapons. Just as nuclear weapons require fissile material, radioactive material is necessary for a radiological dispersal device (RDD). An RDD would need one or more radioactive isotopes, such as cesium-137, strontium-90, or cobalt-60, in combination with an explosive detonation [34]. Nuclear reactors are the sole producer of these three isotopes and many others that an actor might consider using in a RDD. Additive manufacturing alone currently poses no advantage in constructing nuclear reactor components. There exists no substitute for the weapons material itself, therefore additive manufacturing adds little in its acquisition.

The risk associated with each sensitive nuclear and radiological technology is summarized in Table 1. It illustrates that AM adds no risk in obtaining radiological or nuclear material itself. Most technologies fall under the low risk category due to handling of toxic gases or the need to constrain materials in vacuum. The simplest pieces of equipment (end caps, casing, etc.) pose the greatest threat of any technology in the table due to the ability of AM to build pieces with precise specification without excessive bulkiness of the objects. The relative utility of making these pieces with AM has the potential to be marginal, but the flexibility of the machine to make these pieces can increase in the future with suggested improvements in material

Table 1 Risk of additive manufacturing on sensitive nuclear technologies

	Technology	Current	Short-term (5 years)	Long-term (15 years)
Nuclear/Rad Material	Fissile Material (yellowcake, UF6, etc.)	No Risk	No Risk	No Risk
	Radiological (Cs-137, Sr-90, etc.)	No Risk	No Risk	No Risk
Conversion	Steel Vessels	No Risk	No Risk	No Risk
	Gas Storage Tanks	No Risk	Low Risk	Low Risk
	Steel Boiler	No Risk	No Risk	Low Risk
	Drying Kiln	No Risk	Low Risk	Low Risk
	Pumps & Piping	No Risk	Low Risk	Low Risk
Centrifuge Enrichment	Centrifuge Rotors	No Risk	Low Risk	Low Risk
	Casing	Low Risk	Medium Risk	High Risk
	Pumps	Low Risk	Medium Risk	Medium Risk
	Compressors	Low Risk	Low Risk	Medium Risk
	End Caps	Low Risk	Medium Risk	High Risk
	Vacuum Piping	No Risk	Low Risk	Low Risk
	Storage Tanks	No Risk	Low Risk	Low Risk
	UF6 Handling	No Risk	No Risk	Low Risk
Back End/ Reprocessing	Shielding	No Risk	No Risk	No Risk
	Acid Storage Tanks	No Risk	Low Risk	Low Risk
	Centrifuge	No Risk	Low Risk	Low Risk
	Decanter	No Risk	No Risk	Low Risk
	Waste Disposal Systems	No Risk	No Risk	No Risk

Legend

No Risk	(light green)
Low Risk	(green, hatched)
Medium Risk	(red, cross-hatched)
High Risk	(blue, vertical lines)

properties. One could easily produce casing and end caps for centrifuges, as they fit within size constraints, should advance metal AM techniques like laser sintering become commercially available and cost effective.

6 Chemical Applications

The Chemical Weapons Convention identifies three main classes, called Schedules, of controlled substances [35]. Schedule 1 substances have no peaceful use outside chemical weapons while Schedules 2 and 3 substances have small-scale and large-scale uses, respectively, outside chemical weapons. The main substances discussed in this section are sulfur mustard ("mustard gas") and nerve agents, as well as their precursors. Chemical weapons are traditionally difficult to produce due to highly toxic and corrosive chemicals, and their sophistication can vary as evidence of production by the United States, the former Soviet Union, and Iraq.

Sulfur mustard production requires large amounts to be militarily effective. Even if produced in a small quantity by a non-state actor, it is difficult to store and transport. It also possesses a relatively low casualty rate, and medical care has developed to ensure increased recovery rates. Its production historically involves ethylene oxide and hydrogen sulfide [36], both of which are gases at room temperature and therefore difficult to fathom production with AM. The intermediary product between these two chemicals and sulfur mustard is thiodiglycol, which is a common liquid solvent used in ballpoint pen ink and other plastics. It is of interest to private corporations, including Hewlett Packard, who cited it as a functional material in its patent for 3D printing technology in 2017 [37]. This patent does not indicate a threat of thiodiglycol production, but it signals interest of using it by private corporations. Without its direct application, exploration of similar chemicals with 3D printing could generate publicly or commercially available knowledge with utilizing it. Therefore, thiodiglycol is a medium risk in the long term, indicated in Table 2. Thiodiglycol requires

Table 2 Risk of additive manufacturing on sensitive chemical and biological material

	Technology	Current	Short-term (5 years)	Long-term (15 years)
Sulfur Mustard	Thiodiglycol	No Risk	Low Risk	Medium Risk
	Thionyl chloride	No Risk	No Risk	No Risk
	Sodium sulfide	No Risk	No Risk	No Risk
Tabun	Phosphorus oxychloride	No Risk	No Risk	No Risk
	Dimethylamine	No Risk	No Risk	No Risk
	Sodium cyanide	No Risk	No Risk	No Risk
Sarin/Soman	Dimethyl methylphosphonate	No Risk	No Risk	No Risk
	Dimethyl hydrochloride	No Risk	No Risk	Low Risk
	Potassium bifluoride	No Risk	No Risk	No Risk
	Diethyl phosphite	No Risk	No Risk	No Risk
Biological	Seed cultures	No Risk	No Risk	No Risk
	Fermenters	No Risk	Low Risk	Low Risk
	Propagation vessels	No Risk	Low Risk	Low Risk
	Microencapsulation systems	No Risk	No Risk	No Risk

Legend

No Risk	(light green)
Low Risk	(green hatched)
Medium Risk	(red grid)
High Risk	(blue vertical lines)

hydrogen sulfide to produce the sulfur mustard, therefore proliferators need additional anti-corrosive equipment not aided with the use of 3D printing.

The nerve agent tabun poses a similar challenge as the required hydrogen cyanide reagent is necessary [38]. Sarin and soman, other nerve agents, require hydrochloric acid or hydrogen fluoride, both highly corrosive. 3D-printed containers would not withstand storage or transport of these materials. The AM community would need to experiment more with corrosive reactions on mostly metal materials to ensure advantages over steel pipes and containers. Therefore, materials associated with nerve agent production pose a minimal threat. Table 2 shows the relatively small threat that chemical weapons alone pose.

Moderately advanced chemical and pharmaceutical industries can enable chemical weapons production [39, 40]. Successful acquisition would require conversion of a standard plant to one that could produce chemical weapons. It is therefore possible that AM could be used to create equipment originally intended for a chemical plant that is eventually converted to a chemical weapons facility. An article has proposed effects of current AM technology on the chemical industry to include surgical preparation and drug delivery devices [41], although both are only projected and have not been demonstrated outside of an experimental setting. Many 3D printing applications for chemical application cross into the biomedical and biotechnology arena.

7 Biological Applications

Biological weapons have overlaps with the production of chemical weapons with a few exceptions. One needs to produce a significantly smaller amount of harmful biological material to create the same number of casualties as a chemical weapon. They typically fall into two categories, microbial pathogens or toxins. Most research requires technologically sophisticated facilities capable of examining living organisms at the cell level.

Because of this fact, additive manufacturing adds little to a direct threat from biological weapons. Microbial pathogens such as the causative agents of anthrax, brucellosis, and tularemia, must grow in a controlled environment. Producers of these weapons must ensure sufficient protection of the workers to not infect their own population. Bio-printers are typically designed to work with biocompatible material to make pieces to be inserted in or on the human body [42]. Some research has explored the confinement of small bacteria populations within a hydrogel [43], but interactions between bacteria and conditions that permit growth of large populations is not well understood. Additive manufacturing adds little to the picture if a sophisticated facility with highly trained experts is required to understand the phenomena itself let alone the fabrication of a weapon. Building up to a larger set of facilities to acquire an operational capability is not facilitated with additive manufacturing.

Openly-published literature about bio-printing is important. Greater transparency in the capability reduces the security dilemma of biological research. Research on development of antibiotic-resistant bacteria does not intersect with advances in additive manufacturing. The area of interest for potential disruptions is genetic engineering, which alone has the potential to create "supergerms" that are highly resistant [44], notwithstanding overuse of antibiotics.

Current methods to grow biological weapons material with microorganisms involves a seed culture that is fermented. Although advances could improve on growth of microorganism communities, they are not a substitute for the organic material itself. Fermenters for organic culture growth, typically called bioreactors, are complicated machines that are made of stainless steel [45]. Sizes can range from that of a microbial cell (a few square centimeters) to commercial sizes of hundreds to thousands of liters. Smaller sizes have potential to be manufactured with AM, yet supplemental pieces will also be required. This information on AM threats to biological weapons is included with the chemical weapons in Table 2.

Some recent observations have found that more crude biological weapons may be of interest to non-state actors. The ISIS bombers associated with the Brussels attack in 2016 carried rudimentary bombs filled with animal testicles and fecal matter as a primitive biological weapon [46]. These actors may not be interested in creating complex microbial agents or toxins that require complex facilities. These crude weapons have the potential to be used with relatively simple 3D printed systems used for disguise or transmission.

8 Weapons and Delivery Systems

Although additive manufacturing does not impose as a current threat to CBRN technologies, they could potentially have an effect on supplemental systems to deliver or support these types of weapons. Supplemental weapons and delivery systems, such as aircrafts and ballistic and cruise missiles, remain an important component on nuclear WMD, for example. The discharge of a weapon is as important as the weapon itself; if the weapon cannot be delivered to its target, is it an effective weapon? If not practically used, credibility of the delivery system is necessary for a credible deterrent threat. Actors that seek out WMD are also trying to secure delivery systems for those weapons.

The former United States Office of Technology Assessment (OTA) identified weapons systems as technology necessary for rapid and reliable delivery of WMD [45]. The three most prominent forms of interest are cruise missiles, ballistic missiles, and combat aircraft. These can be used to deliver a nuclear weapon to a target, although in different capacities. Weapons systems are not substitutable for the weapon itself, but proliferation of them could enable a higher demand for the weapon itself. If a country can secure a reliable weapon system, all it needs is the weapon itself for a successful weapon to use strategically. What role could additive manufacturing play in producing these crucial parts?

It is important to note that many weapons systems already exist in countries desiring to proliferate in nuclear weapons. Iran, Iraq, and Libya already possess Su-24 missiles purchased from Russia. Many missiles have been developed and tested since then, including dozens by the Democratic People's Republic of Korea [47]. A more modern reference of some simpler weapons delivery system includes rocket launchers, guided missiles, and explosive technology [48]. Some non-weapons states already possess capabilities that could deliver a nuclear weapon at short-range. For most countries that cannot manufacture the capability domestically, buying them is currently the most economic option. The system is also no substitute for the weapon, therefore the barrier of weapon acquisition still exists.

Non-state actors (NSAs) would have the most to gain from the acquisition of a delivery system. This method has proven effective for obtaining weapons and supplemental systems; ISIS has acquired hundreds of millions of dollars-worth of weapons-related supplies seized from the Iraqi government and Syrian rebels [49]. NSAs have successfully taken control of weapons, and this fact should be applied to the potential of theft of a more advanced weapon system. The acquisition of a weapons system in this manner is independent from additive manufacturing, i.e., AM poses little or no additional threat to weapons proliferation.

Cruise missiles and ballistic missiles are two unmanned vehicles for transporting a WMD. Current efforts to stem the trade of cruise and ballistic missiles include the Missile Technology Control Regime (MTCR). This agreement, consisting of 35 of the world's most developed nations, limits the sale of long-range cruise and ballistic missiles (longer than approximately 300 km). Shorter-range missiles are easily available. A non-state actor would therefore want to print a longer-range missile to potentially reach a neighbor or regional target. The main security threat would be to allied countries of the United States. Could a non-state actor 3D print a missile?

First, acquiring the 3D printer capable of missile component production would be difficulty. It can be assumed that a missile needs to be manufactured out of high-strength, versatile metals. Even the most advanced equipment has trouble creating these ideal metals. At Lawrence Livermore National Laboratory (LLNL), for example, scientists are running into issues with 3D printing of metals using laser powder bed fusion, currently "the dominant method for producing 3D printed metal structures." [50]. The powder bed fusion makes use of laser sintering and beam melting, which remain advanced processes that would be difficult to transfer to less-advanced facilities or poorly equipped actors. Porosity remains an issue for these researchers as they are still trying to understand the science of metal vapor in the process. The scale of their implementation is small, at the millimeter level. Making a full missile solely from AM would be almost an insurmountable technical challenge with today's technology. In addition to problems of timeliness and scale, methods such as power bed fusion leaves objects lacking structural strength and utilize a high power to render a small object [51].

The only institutions capable this far of producing some objects for advanced technological systems are the large American corporations. As mentioned earlier, Raytheon 3D printed a missile, but printing spare parts for the satellites is still on the horizon. SpaceX has recently 3D printed a full SuperDraco rocket engine through laser sintering. However, the material used was a superalloy of Inconel, which is several times more expensive than stainless steel. Obtaining access for strong materials is necessary for a well-designed weapon. But one could claim that a non-state actor only needs a crude weapon to successfully set off a WMD. It will still need to invest in an additive manufacturing system to meet that goal. Conservative estimates of an AM machine cost are around $500,000 [52]. This cost would increase for a potential proliferator given lack of sufficient technical expertise and economic infrastructure to produce necessary components. The Office of Technology Assessment report states that rudimentary weapons systems can be bought for about $1 million in 1993 (over $1.5 million in 2016), and this figure could be even cheaper on the black market. With the current state of the technology, it would be more advantageous to purchase a more reliable and known technology. The OTA report states that "expertise is a main ingredient in developing long-range missiles." Because additive manufacturing is a technology in its early stages of widespread deployment, it is unlikely that a proliferator will want to pursue two challenging technologies of which they lack expertise if a cheaper alternative to the same or superior (to what they could produce) technology is available. That increases the uncertainty of success as well as the time to acquire the technology. Some ballistic missiles even use solid fuel, but it is not likely that a proliferator would attempt to make fuel with a 3D printer (even if the materials were available) because solid fuel adds more technical and practical knowledge to understand how to manage it. Liquid fuel is almost always preferred, and 3D printing has no practical advantage with liquids.

The advantage that missiles have to a non-state actor acquiring a nuclear weapon is that only one may be necessary for success of the actor's mission. Terrorism may be perceived as a victory even if only one city is targeted and successfully damaged. Due to this fact, 3D printing may have a slight edge over traditional manufacturing methods because it is preferable when the objects are in small quantities, e.g., it may be more economical to 3D print only a handful of objects rather than invest in a factory, equipment, and labor to produce just as many objects. As with the up-front investment cost of a factory, 3D printing is not immune to that problem. It can be true that 3D printing is currently more economical for smaller batches of items, but the machine cost is a significant proportion of total cost. The National Institute of Standards and Technology (NIST) review[20] approximates the fraction of cost for the additive manufacturing machine itself is two-thirds or more. For metal 3D printers that could theoretically print missiles, the material costs are also high, as expected for engineered metal powders. That fraction is a large investment for an actor to pursue, especially for a non-industrialized country or NSA lacking experience with the technology.

Aircrafts are another delivery system that could theoretically be used to deliver a nuclear weapon. The tactical advantage they have over missiles is that they are commonly accepted around the world as a means for self-defense; they also have numerous commercial applications. It is difficult to ascertain whether aircrafts are an offensive or defensive weapon or for commercial or military purposes. No universal regulations exist that hinder the purchase of aircraft as with the MTCR for missile trade. They are also generally cheaper to produce and have higher survivability as it is a manned vehicle. 3D printing technology is not up to par with manned aerial vehicles. Earlier this year, Airbus announced the first ever 3D-printed aircraft [53]. It is not made out of metal and operates as an autonomous drone. Because of its light weight, a similar vehicle would not be able to hold a nuclear weapon that weighs at least several hundred kilograms. They might be more effective in dispersing chemical or biological weapons but attaining the aircraft itself would be a larger barrier to overcome in addition to the associated technical challenges.

Dispersing a chemical or biological weapon could use some of the same delivery systems as nuclear weapons. Iran reported that Iraq used a variety of systems to deliver its chemical weapons, including airplanes, artillery shells, and rockets [54]. Older US systems developed before and after WWII, such as cluster bombs and mines, have relatively simple geometries that could be made with AM. Household printers might possess such a capability in the near- to long- term. However, it may be easier for a proliferator to purchase or illicitly obtain a simple device from a country already possessing the demonstrated technology. A non-state actor may be more willing to manufacture it themselves to conceal any trail of evidence.

If some delivery systems could be fabricated with AM, the challenge of tracing back the material to the perpetrator. No 3D printer feed material is strictly regulated by export controls, although some high-grade metals reserved for the most advanced printers are prohibited under agreements like the Nuclear Suppliers Group. The growth of the AM industry and improvements in materials may render it difficult to ascertain a proliferator past the private company. Intended use is not a factor in accepting customers in the 3D printing market.

A non-comprehensive list of delivery systems that could be used to deliver or support CBRN weapons is shown in Table 3. Missiles and rockets, as well as other objects that serve as the host body for the weapons material, remain the most sensitive technologies in the near- and far-term. Electronics and explosives, which serve as necessary supplemental systems to many delivery systems, pose little risk when it comes to AM. Rudimentary electronics can be made or purchased with relative ease, and the expertise is widespread. Little publicly available literature exists on 3D-printed explosives, which would be essential for most nuclear and radiological weapons, including improvised nuclear and radiological dispersal devices. An aerial spray tank seems only applicable to chemical and biological weapons, and corrosiveness of these materials hinders its potential use.

9 Importance of Tacit Knowledge

Some alarmists of the threat of additive manufacturing continue to understate the importance of tacit knowledge in AM. Authors may concede that some of it is necessary but then assume that once a piece is finished, it is ready for deployment. It is important to note that 3D-printed objects require a fair amount of post-processing. Casting and molding the piece may be irrelevant, but objects are rough coming off a 3D printer. A delivery system such as a missile or aircraft needs to be finished properly for aerodynamic considerations. Different aerodynamic properties could decrease the accuracy of the weapon, rendering it useless for an actor's mission of destruction should they choose to target a specific location and not wreak general

Table 3 Risk of additive manufacturing for delivery systems

	Technology	Current	Short-term (5 years)	Long-term (15 years)
Delivery	Electronics	No Risk	Low Risk	Medium Risk
	Rockets/Missile Bodies	Medium Risk	High Risk	High Risk
	Artillery Shells	Low Risk	Medium Risk	Medium Risk
	Cluster Munitions	No Risk	Low Risk	Low Risk
	Aerial Spray Tanks	No Risk	Low Risk	Low Risk
	Combat Aircraft	No Risk	No Risk	No Risk
	UAV	No Risk	Low Risk	Low Risk
	Explosive Materials	No Risk	No Risk	No Risk

Legend

No Risk	(light green)
Low Risk	(green, hatched)
Medium Risk	(red, cross-hatched)
High Risk	(blue, vertical lines)

havoc. Grinding, sanding, and polishing would be skills required to bring the object to its intended use. Expertise in that area is still being developed. This is not to say that finishing a 3D printed object requires a significant amount of effort, but it is important for sensitive weapons systems. Welding is another skill that is necessary for AM applications. If a nation wanted to 3D-print a missile, they are most likely going to have to weld materials together. The 3D printers that can print the ORNL Shelby Cobra or a small plane are the highest quality printers in the US infrastructure. It is unlikely a nation to indigenously manufacture a 3D printer of that quality or to buy it from the United States.

Nuclear weapons have an extensive history of proliferation through spread of tacit knowledge as well as technology. The AQ Khan network remains the most infamous nuclear proliferation networks, which contributed to the nuclear weapons acquisition of North Korea [55]. Scholars have also noted the understatement of tacit knowledge in the spread of biological weapons/terrorism [56]. Some scholars, such as Collins [57] and Vogel [58], have analyzed the theoretical framework of tacit knowledge applied to technology transfer and the application to weapons proliferation, respectively.

Tacit knowledge is similarly important in 3D printing as machine failures and material wastes can be significant without technical experience. The adoption of 3D printing technology is not simply a matter of detailing scientific or engineering advances to a new process." [59]. Tacit knowledge is important in all CBRN weapons development. A prominent example includes safety precautions in order to prevent harm to operators; it is challenging to know problems without having operated the equipment before. Safety concerns are of little importance to 3D printers, but economic considerations are important to ensure functionality of a machine with little technical support for proliferators. Communal tacit knowledge by a small group or larger scientific community may not transfer well to a proliferator that can obtain a sophisticated 3D printer. Tacit knowledge highlights the actualization of weapons-usable material after technological acquisition. Equating the two is an invalid assumption.

10 Conclusions and Recommendations

This work concludes that additive manufacturing overall poses a miniscule threat of WMD acquisition of rogue states or non-state actors. The CBRN material cannot be produced with 3D printers alone, and the mechanical and chemical processes are challenging with today's technology. Little signs point to a disruptive capability of AM to proliferate WMD to aspirant proliferators. Delivery systems remain the most prevalent opportunity for proliferators to use AM; small yet complex objects like casing or bodies of these systems are ideal candidates for AM pieces. It is important as well to view weapons acquisition with AM through the lens of relative gains compared to traditional manufacturing methods or other means of technological acquisition; proliferators could look to AM or other similar technologies as covert, innovative, and cost-effective ways to increase their power and leverage.

Because additive manufacturing is not on the brink of threatening international stability, it would be wise to monitor its progress in the near future. Although 3D-printed missiles or aircraft capable of delivering WMD may not be used next year, the industry is growing rapidly. Creating a governmental entity or committee that monitors the capabilities of additive manufacturing in the United States and worldwide is recommended. Qualified technical and security experts and policymakers should be on such a body, and it could potentially make recommendations on sensitive technologies when they are transferred. This body would work closely with the Nuclear Suppliers Group and other international trade groups/agreements to update the agreements to include AM. Currently, there is not a strong need to strictly limit the technology, but with more advances in sensitive areas, AM should be viewed as a dual-use technology. Although detection will be challenging, export controls will need to be enacted to ensure proper end use of the technology.

Due to the potential transferability of files, cybersecurity should be strengthened of organizations, such as defense contractors, that may use this technology for military applications. Understanding of computer design programs is more widespread, and it would be easier for a relatively unskilled actor to print an object from a 3D file. The body mentioned above could also attempt to develop a classification system for different files, e.g., if it is a handgun or a toy or a piece for an engine or motor. An attempt to gather intent from the files, although challenging, could be useful monitoring tool and could aid in classifying facilities based on files they download or produce.

It would also be wise to limit the domestic use of AM for sensitive technologies or to split into multiple files. Saving a 3D file for a centrifuge, for example, is too risky to maintain on a single file. Even if an actor could not 3D print the piece, insight can be gained from the file itself, e.g., specific geometries or supplemental systems. It is possible to entertain the idea of making some of the manufacturing techniques confidential so as the spread of this eventual dual-use technology is curtailed. This action could also aid the US economy should AM become a viable large-scale production method.

The US should also begin to consider attribution methods should a weapon be used with AM technology. Being able to distinguish the source of the material helps the problem of perpetrator identification. Effort has been placed in a similar area of nuclear forensics, where it is possible to ascertain the origin of source material given concentrations of some trace isotopes. Attempting a similar approach with this technology would be challenging as the material is more ubiquitous and uniform, therefore extensive international and public/private efforts should be investigated. Investing in methods to determine where a weapon originated could help deter some actors from pursuing weapons with AM technology.

Monitoring the progress in conjunction with other technologies, primarily genetic engineering and nanotechnology, is also important for the future. Genetic engineering could manipulate organisms to have traits lending to increased survivability. Cultivation would then be the primary hurdle to biological weapons acquisition. Nanotechnology, on the other hand, has potential to improve material properties that are insufficient for full AM-WMD production today. Nonetheless, additive manufacturing technology alone remains inadequate to translate directly to a WMD capability in the short term. Advances in its time, cost, and material properties as well as complementary technologies could attract proliferators to implement in some capacity in the longer term.

References

1. Mattox JM (2013) Additive manufacturing and its implications for military ethics. J Mil Ethics 10(3):225–234
2. Schwab K (2016) The Fourth Industrial Revolution: what it means, how to respond. World Economic Forum, 14 Jan 2016. https://www.weforum.org/agenda/2016/01/the-fourth-industrial-revolution-what-it-means-and-how-to-respond/
3. Bloem J et al (2014) The Fourth Industrial Revolution: things to tighten the link between IT OT. VINT Research Report, Sogeti. https://www.fr.sogeti.com/globalassets/global/downloads/reports/vint-research-3-the-fourth-industrial-revolution
4. Petrick IJ, Simpson TW (2015) 3D printing disrupts manufacturing: how economies of one create new rules of competition. Res-Technol Manag 56(6):12–16

5. Campbell TA, Ivanova OS (2013) Additive manufacturing as a disruptive technology: implications of three-dimensional printing. Technol Innov 15(1):67–79
6. Price compare—3D printing materials—Filament Finder. 3D printing from scratch (2019). http://3dprintingfromscratch.com/3d-printer-filament-finder/
7. 3D Printing Materials: Printing Sizes. i.materialize (2019). https://i.materialise.com/3d-printing-materials/printing-sizes
8. Thingiverse (2019). http://www.thingiverse.com/. Accessed 14 Feb 2019
9. YouMagine (2019). https://www.youmagine.com/. Accessed 14 Feb 2019
10. Tinkercad. https://www.tinkercad.com/. Accessed 14 Feb 2019
11. Jackson R et al (2016) Overview of the Oak Ridge National Laboratory Advanced Manufacturing Integrated Energy Demonstration Project: case study of additive manufacturing as a tool to enable rapid innovation in integrated energy systems. ASME 2016 international mechanical engineering congress and exposition, vol 14, Nov 2016
12. Werner D (2018) Lockheed Martin extends additive manufacturing to key spacecraft components. Space News (2018). https://spacenews.com/lockheed-martin-extends-additive-manufacturing-to-key-spacecraft-components/
13. To Print a Missile: Raytheon Research Points to 3-D Printing for Tomorrow's Technology. Raytheon, Apr 2016. http://www.raytheon.com/news/feature/3d_printing.html
14. 3D Printed Shelby Cobra. Oak Ridge National Laboratory. http://web.ornl.gov/sci/manufacturing/shelby/
15. Feuer A (2015) Cody Wilson, who posted gun instructions online, Sues State Department. The New York Times. http://www.nytimes.com/2015/05/07/us/cody-wilson-who-posted-gun-instructions-online-sues-state-department.html
16. Defense Distributed vs. United States Department of State. Fifth Circuit 2016, Harvard Law Review, Apr 2017. https://harvardlawreview.org/2017/04/defense-distributed-v-united-states-department-of-state/
17. Jain PK, Pandey PM, Rao PVM (2009) Effect of delay time on part strength in selective laser sintering. Int J Adv Manuf Tech 43(2009):117–126
18. The US National Security Strategy. The White House, Feb 2015. https://obamawhitehouse.archives.gov/sites/default/files/docs/2015_national_security_strategy_2.pdf
19. Kroenig M, Volpe T (2015) 3-D printing the bomb? The nuclear nonproliferation challenge. Wash Q 38(3):7–19
20. How 3D Printers Work. US Department of Energy. http://www.energy.gov/articles/how-3d-printers-work. Accessed 19 June 2014
21. Gress D, Kalafsky R (2015) Geographies of production in 3D: theoretical and research implications stemming from additive manufacturing. Geoforum 60:43–52
22. Walther G (2015) Printing insecurity? The security implications of 3D-printing weapons. J Sci Eng Ethics 21:1435–1445. https://doi.org/10.1007/s11948-014-9617-x
23. McNulty M et al (2012) Toward the printed world: additive manufacturing and implications for national security. Institute of National Strategic Studies, National Defense University, Sept 2012
24. Lothringer D et al Countering weapons of mass destruction: a preliminary field study in improving collaboration. Naval Postgraduate School, Mar 2016
25. Huang Y et al (2015) Additive manufacturing: current state, future potential, gaps and needs, and recommendation. J Manuf Sci Eng 137(1):014001–014001-10
26. Tirone D, Gilley J (2015) Printing power: 3-D printing and threats to state security. J Polic Intell Count Terror 10(2):102–119
27. Walther G (2015) Printing insecurity? The security implications of 3D-printing weapons. J Sci Eng Ethics 21:1435–1445
28. Albright D (2018) Technical note: making sense out of the IR-8 centrifuge. Institute for Science and International Security, 23 Sept 2018. http://isis-online.org/uploads/isis-reports/documents/IR8_Sept__2014.pdf
29. Guidelines for Nuclear Transfers (2015) Nuclear Suppliers Group, June 2015. http://www.nuclearsuppliersgroup.org/images/Files/Updated_control_lists/Bariloche/NSG_Part1Rev.13_clean.pdf

30. Manfredi D et al (2012) From powders to dense metal parts: characterization of a commercial AlSiMg allow processed through direct metal laser sintering. Materials 6(3):856–869
31. Albright D, Hinderstein C (2010) Unraveling the AQ Khan and future proliferation networks. Wash Q 28(2):109–128
32. Natanz Enrichment Complex. Nuclear threat initiative. http://www.nti.org/learn/facilities/170/. Accessed 7 July 2017
33. Zentner MD, Coles GA, Talbert RJ (2005) Nuclear proliferation technology trends analysis. US Department of Energy, Office of Scientific and Technical Information, Pacific Northwest National Laboratory, Sept 2005. https://www.osti.gov/biblio/885472-nuclear-proliferation-technology-trends-analysis
34. Hautecouverte B (2009) A possible international regime to cover radiological materials. Centre d'Études de Sécurité Internationale et de Maîtrise des armements. https://inis.iaea.org/collection/NCLCollectionStore/_Public/48/018/48018671.pdf
35. Annex on Chemicals. Organisation for the Prohibition of Chemical Weapons. https://www.opcw.org/chemical-weapons-convention/annexes/annex-chemicals/annex-chemicals
36. Toxicological Profile for Sulfur Mustard. Agency for Toxic Substances & Disease Registry, Centers for Disease Control, Sept 2003. https://www.atsdr.cdc.gov/toxprofiles/tp49.pdf
37. Kabalnow A et al (2017) Three-dimensional (3D) printing. Hewlett Packard Development Company, L.C. WO Patent App. PCT/US2015/041,961, Feb 2017
38. Technologies Underlying Weapons of Mass Destruction (1993) US Office of Technology Assessment, OTA-BP-ISC-115. US Government Printing Office, Washington, D.C.
39. Chemical and Biological Weapons: The Poor Man's Bomb. Federation of American Scientists, Intelligence Resource Program, Oct 1996. https://fas.org/irp/threat/an253stc.htm
40. Tucker J (1998) Verification provisions of the chemical weapons convention and their relevance to the biological weapons convention. In: Biological weapons proliferation: reasons for concern, courses of action. The Henry L. Stimson Center, pp 76–104
41. Gross B et al (2014) Evaluation of 3D printing and its potential impact on biotechnology and the chemical sciences. Anal Chem 86(7):3240–3253. https://pubs.acs.org/doi/10.1021/ac403397r
42. Nishiyama Y et al (2008) Development of a three-dimensional bioprinter: construction of cell supporting structures using hydrogel and state-of-the-art inkjet technology. J Biomech Eng 131(3):035001-6. https://doi.org/10.1115/1.3002759
43. Connell J, Ritschdorff ET, Whiteley M, Shear JB (2013) 3D printing of microscopic bacterial communities. Proc Natl Acad Sci (PNAS) 110(46):18380–18385
44. Cheng Y-A, Li M (2010) Get to know supergerms better. Chin J Dermatovenerol 24:1081–1083
45. Technologies Underlying Weapons of Mass Destruction. US Congress, Office of Technology Assessment, OTA-BP-ISC-115. US Government Printing Office, Washington, D.C., Dec 1993
46. Alle P (2016) Brussels terror suspect who was shot by police at tram stop had rucksack containing animal testicles and faeces, prompting fears ISIS is planning crude biological attack on food supplies. Daily Mail. http://www.dailymail.co.uk/news/article-3529865/Brussels-terror-suspect-shot-police-tram-stop-rucksack-containing-animal-testicles-faeces-prompting-fears-ISIS-planning-crude-biological-attack-food-supplies.html
47. Berlinger J (2017) North Korea's missile tests: what you need to know. CNN, 3 Dec 2017. http://www.cnn.com/2017/05/29/asia/north-korea-missile-tests/index.html
48. Armament Technologies. Defence Research and Development Organisation, Aug 2012
49. Laub Z (2016) The Islamic State. Council on Foreign Relations. https://www.cfr.org/backgrounder/islamic-state. Accessed 10 Aug 2016
50. Matthews M et al (2016) Denudation of metal powder layers in laser powder bed fusion process. Acta Meterialia 114(1):33–42
51. Power Bed Fusion. Additive Manufacturing Research Group, Loughborough University (2019). http://www.lboro.ac.uk/research/amrg/about/the7categoriesofadditivemanufacturing/powderbedfusion/
52. Thomas D, Gilbert S (2014) Costs and cost effectiveness of additive manufacturing: a literature review and discussion. Applied Economics Office Engineering Laboratory, National Institute of Standards and Technology, Dec 2014. http://dx.doi.org/10.6028/NIST.SP.1176

53. Shivali B (2016) World's first 3D printed plane unveiled: airbus' windowless 'Thor' aircraft could pave the way for cheaper and faster flights. Daily Mail. http://www.dailymail.co.uk/sciencetech/article-3627187/World-s-3D-printed-plane-unveiled-Airbus-windowless-Thor-aircraft-pave-way-cheaper-faster-flights.html. Accessed 6 June 2016
54. Medical Aspects of Chemical and Biological Warfare. Editor in Chief, Brigadier General Russ Zajtchuk. Office of The Surgeon General Department of the Army, Bethesda, United States of America (1997)
55. Montgomery A (2005) Ringing in proliferation: how to dismantle an atomic bomb network. Int Secur 30(2):153–187
56. Revill J, Jefferson C (2014) Tacit knowledge and the biological weapons regime. Sci Public Policy 41(5):597–610
57. Collins H (2012) Expert systems and the science of knowledge. In: Bijker WE, Hughes TP, Pinch T (eds) The social construction of technological systems: new directions in the sociology and history of technology. MIT Press, Boston, pp 321–348
58. Vogel K (2006) Bioweapons proliferation: where science studies and public policy collide. Soc Stud Sci 36(5):659–690. 706059460
59. Hoskins S (2014) Issues of tacit knowledge within 3D printing for artists, designers and makers. In: NIP & digital fabrication conference, vol 30, pp 426–431. http://eprints.uwe.ac.uk/id/eprint/23449

Machine Learning in the Countering Weapons of Mass Destruction Fight

Peter R. Exline

Abstract Conflict between states in the modern era takes place under the threat of nuclear weapons use. Preventing additional states, especially adversarial ones, from acquiring nuclear weapons is the goal of the United States Department of Defense's Countering Weapons of Mass Destruction (C-WMD) program as defined in Joint Publication 3-40. This chapter analyzes the utility of machine learning in assessing specific indicators of nuclear proliferation based on feasibility and utility criteria. Nuclear proliferation indicators are developed and machine learning evaluation criteria designated and discussed. Implications for chemical and biological weapons are briefly discussed. A speculative look at far-future, true generalized artificial intelligence in the C-WMD fight is made, with a focus on determining new questions that could be answered by an advanced system. The results show that the most promising areas for machine learning in Counter-WMD are power grid analysis, imagery analysis to located hidden and protected sites, and communications metadata analysis to identify key players and their activity in proliferation networks. Far-future artificial intelligence may be able to track proliferator progress, anticipate nuclear decision points, and design new arms reduction frameworks.

1 Introduction

The goal of this chapter is to provide an initial assessment of the feasibility and utility of applying machine learning to Counter-WMD efforts, especially in the nuclear arena. As a prelude to analysis, intelligence indicators of proliferation activity will be developed. US Joint Publication 2-0, *Joint Intelligence*, defines indicators as follows: "In intelligence usage, an item of information which reflects the intention or capability of an adversary to adopt or reject a course of action." [1].

These indicators will serve as independent variables, seeking to answer the question: "If machine learning is used to analyze indicator X, will it provide output

P. R. Exline (✉)
Georgia Institute of Technology, Atlanta, GA, USA
e-mail: pete.exline@gatech.edu

M. E. Kosal (ed.), *Disruptive and Game Changing Technologies in Modern Warfare*, Advanced Sciences and Technologies for Security Applications,
https://doi.org/10.1007/978-3-030-28342-1_5

exceeding the utility of human analysts?" Additional nuance is developed for the dependent variable, "utility of human analysts" in the subsection, Machine Learning Feasibility Criteria, which analyzes relative value and near future potential. In a separate section, possible distant future impacts on detection of WMD proliferation from generalized AI are discussed. The working hypothesis of this research is that machine learning used to analyze Counter-WMD indicators will only provide output exceeding the utility of human analysts in select areas with marginal gains.

2 Indicators and Feasibility Criteria

2.1 Nuclear Proliferation Indicators

There are four major hurdles to creating a nuclear weapon according to the US Department of Defense (DoD) [2]. Fissile material must be made or acquired. A workable design must be researched or obtained, and the device manufactured. Finally, the device must be weaponized for delivery via bomber or rocket. Of these steps, obtaining the fissile material is the most difficult and the focus of Counter-WMD efforts. The other steps in creating a nuclear weapon will be discussed first, and obtaining fissile material is covered in detail in a subsequent section.

2.1.1 Weapon Design

An operable nuclear device requires sufficient fissile material to form a supercritical mass leading to a runaway nuclear reaction causing a nuclear explosion. When fission begins, the mass tends to separate due to the energy release and some method of maintaining sufficient density, whether through geometry, compression, or a combination thereof, must be built.

There are two primary designs of creating a sustainable weapon for delivery of a nuclear bomb: a gun-type device and an implosion device. The *Nuclear Matters Handbook* illustrates both as shown in Figs. 1 and 2.

The gun-type device is simpler to design and assemble and more likely to result in a nuclear explosion if the proliferator lacks technical expertise and tacit knowledge required for an implosion device [3]. It operates by splitting the fissile material into separate pieces that are individually subcritical. A chemical explosive then propels one piece into the other, creating the required supercritical assembly.

The gun-type assembly was considered foolproof and in 1945 the US detonated this type of assembly, also known as the Little Boy fission weapon, over Hiroshima without prior testing [4]. The simplicity of this weapon delivery system comes with drawbacks, however. It is less efficient in terms of yield vs. required fissile material than the implosion device, which results in a larger and heavier weapon, though harder to deliver. While the gun-type device sets the minimum requirements that a

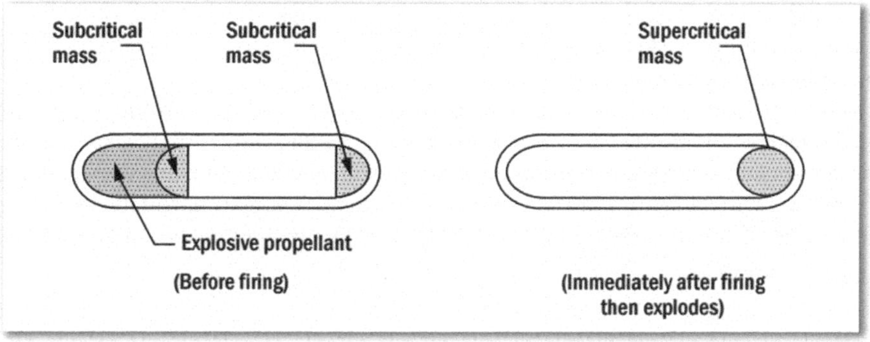

Fig. 1 Unclassified illustration of a gun-type device. (*Note* The appearance of US Department of Defense (DoD) visual information does not imply or constitute DoD endorsement) [3]

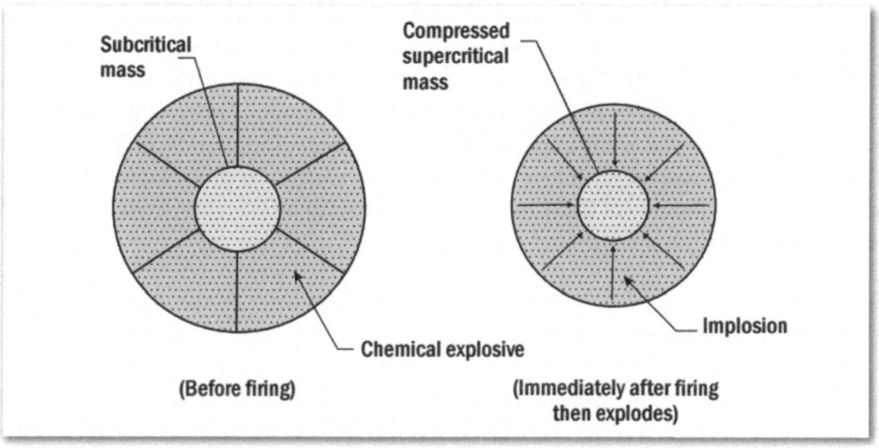

Fig. 2 Unclassified illustration of an implosion device. (*Note* The appearance of US Department of Defense (DoD) visual information does not imply or constitute DoD endorsement) [3]

non-state actor might pursue for nuclear terrorism, a state proliferator will likely opt for the implosion device to gain larger yields or more weapons from limited fissile material.

The implosion device is more technically complex. A subcritical amount of fissile material is assembled into a sphere. Chemical explosives are placed around it and, when triggered, compress the subcritical mass into a density that makes it supercritical [5]. Designing the chemical explosives and firing mechanism in this implosion device is non-trivial. However, AQ Khan, a key figure of the Pakistani nuclear weapons development program, shared a successful design with several nations including Iran, North Korea, and Libya [6]. A state with interests in developing nuclear weapons with the resources to obtain fissile material would likely be able to purchase, steal, or otherwise obtain such a design, overcoming a central barrier to nuclear proliferation.

Due to the likelihood of an implosion design being used by a proliferating state [7], indicators of proliferation are generated for purposes of this research study. These indicators include (1) hiring or development of chemical explosives experts, (2) import or manufacture of the explosives in quantities/types unsuitable for other military applications, and (3) electronic expertise/equipment suited uniquely to an implosion weapon.

2.1.2 Manufacture

The tools needed to manufacture a modern fission weapon are technically advanced and involve specific requirements. Many of these items are considered dual-use—they have legitimate, non-nuclear-weapons uses, including medical and industrial applications. The key equipment required, both dual-use and nuclear-specific, is tracked by the Nuclear Suppliers Group, an organization consisting of states which export materials for use in nuclear power plants and weapons programs, and which develops standards for such exports [8]. The Nuclear Suppliers Group and their efforts will be discussed further in the "Analysis" section of this chapter. Tracking purchases and shipments of these items provides useful information in detecting and tracking proliferation attempts.

2.1.3 Weaponization

Weaponization of the nuclear device requires it to be miniaturized and ruggedized so that it can be delivered by a traditional military system, such as a bomber or a rocket [7]. While not a requirement for nuclear terrorism, a state actor would seek to weaponize their nuclear devices to either attack an enemy state or for deterrent effects. With the advent of modern air defense systems, a state using a bomb design would also need a survivable air platform, such as a modern jet—further driving size and weight reductions, and requiring more testing or acquisition of advanced designs. Weight and size requirements for rocket delivery are stringent and the device must withstand a much higher acceleration than would be necessary for bomb delivery.

Weaponization is achieved through iterative testing (including full-scale nuclear tests) or acquiring known designs from nuclear states. Further, the delivery vehicles themselves must be acquired or built. Space programs are often dual-use, as a rocket capable of reaching Earth's orbit with a payload intact is very useful as a nuclear weapon delivery system. Thus, this process presents several types of indicators of a nuclear program and its level of progress.

2.1.4 Fissile Material

Obtaining sufficient quantities of fissile material is the most difficult part of the nuclear proliferation process. There are two primary candidates: uranium-235 (^{235}U) and plutonium-239 (^{239}Pu). The International Atomic Energy Agency (IAEA) defines "significant quantities" of ^{235}U as 25kg (>20% enrichment; i.e., Uranium with greater than 20% ^{235}U) and 8 kg of Pu (<20% ^{238}Pu, normally achieved by increased ^{239}Pu content) [9]. These are general benchmarks on the level of uranium or plutonium enrichment that would be required for creating a nuclear device.

2.1.5 Enrichment of Uranium

Natural uranium consists of approximately 99.28% ^{238}U and only 0.72% ^{235}U [10]. ^{238}U can be used to create useful plutonium isotopes, but it is not considered fissile material, or that which could be immediately used for a nuclear device. If enriched above 3% ^{235}U, the uranium is useful as civilian nuclear reactor fuel. Up to 20% enrichment may be justified for research reactors and medical isotope production. Enrichment of ^{235}U beyond 20%, or a large quantity enriched to levels approaching 20%, are considered indicators of a nuclear weapons enrichment process [11].

For a state developing the technology to produce nuclear weapons, there are three main approaches to enrichment: gas centrifuges, gaseous diffusion, and electromagnetic separation.[1] All three approaches take advantage of the slight difference in mass of the isotopes in different ways. Centrifugal separation spins the isotopes (gas form of UF_6) and leverages the heavier isotopes being flung outward further than the lighter ones. Gaseous diffusion converts the material to gas form (usually UF_6) and uses the relatively higher velocity of lighter isotopes at the same kinetic energy as heavier isotopes, leading to faster diffusion through membranes [14]. Finally, electromagnetic separation uses the material in ion form, passing the material through a directional magnetic field, and using mechanical physics to create a different radius of curvature, where separate collection points exist for the light and heavy isotopes.

Regardless of the method selected, two key indicators that reflect nuclear enrichment are created: a large amount of space is needed for enrichment and the energy requirements are very high. The latter is one of the easiest indicators of enrichment activities from an intelligence standpoint if the facility is isolated from populated or heavily industrialized areas [15]. A nuclear reactor itself is not well-suited to direct ^{235}U production, though it could in theory power another enrichment process.

The IAEA has an international inspection regime in place for approved uranium enrichment facilities [16]. However, as the inspection arm of an international treaty, the IAEA cannot truly enforce agreements, nor can they inspect sites to which the

[1]This reference discusses gas centrifuges and gaseous diffusion: [12]. Additionally, electromagnetic separation was used at the Oak Ridge site by the United States during World War II, per L. Love [13].

host state denies them access. A state refusing IAEA inspections may also be an indicator of a nuclear weapons program.

Inspectors encounter difficulty in assessing uranium enrichment even when international inspections are allowed. It can be difficult to determine the order in which gas flows through the cascade of centrifuges. If an operator changes valve settings, the same array of centrifuges could be used to enrich large amounts of slightly enriched uranium for peaceful uses or small amounts of highly enriched uranium for weapons use [17]. Only truly invasive inspections can ascertain the purpose of enrichment, and even then the functions can be switched between inspections, especially if the inspections are infrequent or require significant prior warning.

2.1.6 Enrichment of Plutonium

There are several fissile isotopes of plutonium, but the most commonly pursued is ^{239}Pu. The most straightforward method of producing this isotope is bombarding natural uranium (primarily ^{238}U) with neutrons for a set amount of time, then removing the uranium and chemically extracting the plutonium. Some commercial reactors are capable of producing ^{239}Pu if the amount of time the uranium is exposed to neutrons is controlled—normally a much shorter period of time than a typical nuclear reactor fuel cycle.

The amount of bombardment time must be limited because, while increased time will increase ^{239}Pu amounts, undesirable isotopes are created [18], including ^{240}Pu which has a high spontaneous fission rate. Spontaneous fission is undesirable in a nuclear weapon because this could lead to premature denotation before the desired supercriticality is reached for higher yield devices, leading to inconsistent performance and or duds. Further, such spontaneous fission emits dangerous radiation and make the material dangerous to handle.

The differences in enrichment methods for uranium and plutonium result in an interesting decision for the proliferating state. Plutonium has the advantage of a relatively straight-forward process that leverages a dual-use civilian nuclear energy process for plausible deniability and obscuration of a nuclear weapons program. However, if the state is suspected of developing nuclear weapons, the nuclear reactor presents a large, vulnerable, and expensive fixed target for enemies to attack. Conversely, uranium enrichment is easier to hide and disperse, but such enrichment in quantities exceeding the minimal amount necessary for nuclear power reactors are a very strong indicator of a nuclear weapons program and can lead to international isolation and intervention. Counter-WMD efforts must account for each approach.

2.1.7 Summary of Indicators of Nuclear Proliferation

The analysis thus far has focused on the physical and design requirements for nuclear weapons programs. In Fig. 3, the "Material Acquisition" step is the most difficult to

Fig. 3 Nuclear event pathway. US Department of Defense (*Note* The appearance of U.S. Department of Defense (DoD) visual information does not imply or constitute DoD endorsement) [19]

accomplish for a proliferator, and is also the step that may give the largest, clearest indicators of nuclear proliferation.

The "Motivation and Planning" step is beyond the scope of this research as the rationale for pursing nuclear weapons range from regime preservation to mutual deterrence to state prestige [20]. Traditional methods and metrics from political science and international relations can be used to estimate likelihood of this step being pursued and completed.

In the "Material Acquisition" step, indicators of a nuclear weapons program include:

- Acquisition of raw (unenriched) special nuclear material precursors
- Enrichment of uranium in quantities exceeding civilian nuclear energy needs
- Enrichment of uranium to levels beyond the 20% necessary for research reactors or medical isotope production
- Enrichment of plutonium indicated by artificially short fuel burn time in civilian nuclear reactors
- Excess power generation/distribution in areas without a clear need (powering enrichment processes covertly)
- Evidence of large scale excavations and other traditional intelligence indicators of hidden and/or protected facilities
- Refusal to give access to international inspectors or limitations on sites to be inspected.

In the "Design & Fabrication" step, indicators include:

- Hiring or training of nuclear weapons scientists
- Seeking out or acquisition of existing nuclear weapons designs (including the likes of the A.Q. Khan network)
- Acquisition of specific or dual-use equipment required for fabrication and tracked by the Nuclear Suppliers Group
- Acquisition or manufacture of chemical explosives in excess of conventional military or civilian requirements
- Acquisition of complex electronics necessary to create a firing mechanism of an implosion device
- Development of delivery systems, including delivery systems used in peaceful space programs
- Evidence of nuclear testing, including underground or undersea.

In the "Storage and Movement" step, efforts by agencies such as the Defense Threat Reduction Agency, the Department of Homeland Security, the US Coast Guard, and foreign partners seek to identify movement of materials or weapons through detectors and inspections of moving cargo.

The indicators listed above, while not an exhaustive list, will be analyzed in this chapter in an attempt to determine how well machine learning techniques and systems can process potential indicators of nuclear weapons programs and inferences drawn as to the whether this capability surpasses the capability and speed of human analysts.

2.2 *Chemical and Biological Proliferation Indicators*

Chemical and biological weapons do not have a singular "long pole in the tent" similar to nuclear weapons programs, which have a strict requirement for fissile material, generally obtainable in only two ways, both of which have specific indicators. Further, chemical, and to a lesser extent biological, weapons programs can be disguised in true dual-use facilities for the creation, storage, and even testing of their use [20]. As such, rather than attempt to track these materials, machine learning might focus on evidence of hidden facilities or pursuit of information/expertise specific to weapons use. One indicator that may be more prevalent for chemical and biological weapons would be the acquisition of appropriate protective gear and other countermeasures. Another is the development of low-to-mid-range delivery systems capable of carrying chemical or biological warheads, though these may be difficult to distinguish from systems for normal explosives.

3 Machine Learning Feasibility Criteria

Machine learning leverages large computational power to quickly analyze large amounts of data to produce useful information. While the theory and approaches are decades old, only in more recent years has sufficient computer power become available to make it useful in solving large and complex problems.

One of the most remarkable successes of machine learning was the defeat of chess grandmaster Garry Kasparov at the hands of IBMs' Deep Blue chess computer in 1997 [21]. Kasparov, the reigning world champion at the time, had defeated Deep Blue a year earlier, losing the first match, but developing strategies that exploited weaknesses in the machine learning's approach to come back and win the series. In the ensuing year, the algorithms were updated and more computational power added, leading to Deep Blue's win in the series that came down to the final match [22].

The game of chess has a tremendous decision space which covers over 10^{123} possibilities—many orders of magnitude larger than the approximate number of atoms in the universe (~10^{80}) [23]! Solving this problem through brute force—calculating each possible game and selecting optimally—is literally impossible given the constraints of our universe. Therefore, a computer playing chess must take another approach. The first approaches were in the nascent artificial intelligence field. Programmers sought to teach a computer to "think" like a human would when playing chess. Much like humans' attempt to fly by mimicking birds failed, so did this approach [22].

Although it frustrated artificial intelligence purists [22, p. 3], efforts shifted to algorithmic approaches that combined human understanding of chess logic, moves, and outcomes with the computer's ability to rapidly evaluate the utility of future moves. Without the human design and input of evaluation criteria, the effort at machine learning would be far less efficient. However, over the course of multiple games or data sets, the machine can "learn" patterns and optimizations of its own. This machine learning process does not take place in the same way human learning does, but it is highly effective.

3.1 Subsets of Machine Learning

The combination of machine iterative power, defined inputs, defined outputs and a task to map one to the other with defined evaluation criteria is a subset of machine learning known as supervised learning [24]. Contrast this to unsupervised learning, where the outputs are explicitly defined and the machine looks for regularities in inputs. Such an approach is used in determining Optical Character Recognition algorithms, for instance [24]. In other problems, the individual actions from input to output are not important, whereas the value of the destination (or total path) matters. Reinforcement learning is the machine learning subset used in this research study [24].

3.2 Limitations of Machine Learning

There is an inductive bias for traditional machine learning approaches, shown in Fig. 4, which uses an intentionally contrived artificial set of data points intentionally chosen to illustrate the concept. Depending on the human programmer's assumption about the data, either of the two lines drawn may be the "best fit" for the data.

If the original origin and specific accuracy and precision in measuring the date—what might be colloquially referred to as the data's "true" nature—is unknown, there is a risk of overfitting the data. The blue line in Fig. 4 represents a higher-order polynomial fit to the data. If the data is not truly a high-order polynomial function, the numerous inflections on the blue line are spurious and due to noise in the data, and have no predictive value. But if the machine is programmed to fit the data to a high-order polynomial—an underlying assumption—this will be used. Taken to an extreme, a really-high-order polynomial could literally draw a line through each data point—but the resulting line would have no value at all.

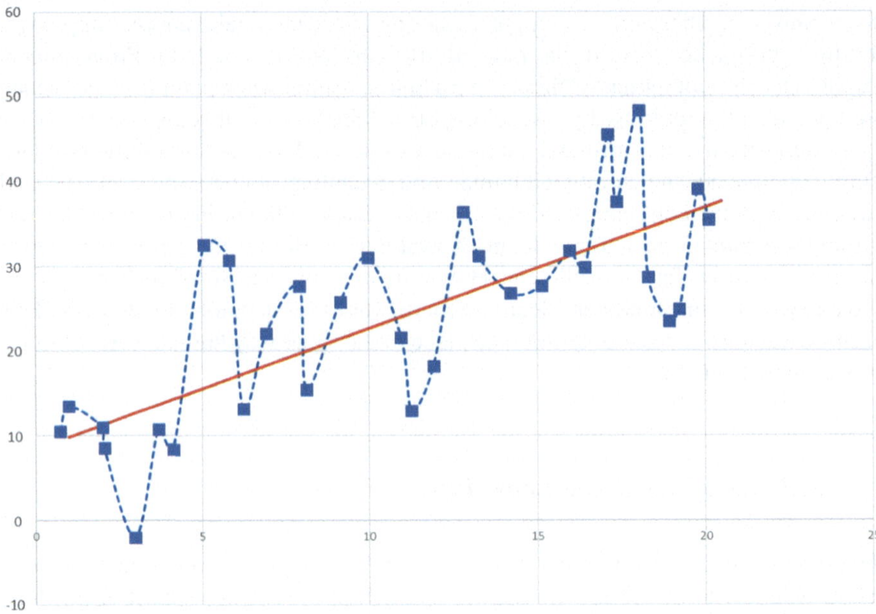

Fig. 4 Different fits for the same data

Conversely, underfitting the data during the machine learning process is also a risk. If the phenomena are assumed to be linear, a line such as the red line in Fig. 4 results. Real cause and effect relationships in the data are disregarded and are assumed to be noise, again destroying the predictive value of the machine learning process.

While in this simple example, both approaches at fitting the data could be attempted and evaluated against each other, this type of evaluation could also result in inductive bias due to the humans setting the evaluation criteria.

Machine learning applications do best when the data input aligns with familiar or training data, as machines often struggle initially with new data [25]. Further, optimizations that help with specific applications, such as facial recognition, reduce the machine's ability to solve more general problems [26]. An attempt to improve on this conundrum is "deep learning," whereby the machine first looks for simple patterns and then builds on layers of complexity. However, there are also issues with this approach as the problem is only abstracted a few layers. The machine still needs large data sets with human-labelled training data and human evaluation criteria at the end. The machine does not truly "think" or "understand", though such deep learning certainly has utility—rapid identification of objects in images can save countless hours of human analyst work. This approach works best with large data sets and can handle unknown inputs, but fails when the number of possible models is large [27].

4 Evaluating Machine Learning for Security Applications

4.1 Background

The effective use of machine learning in a military context is not science fiction. The Swedish Defense Department used machine learning to analyze submarine incursions into its territorial waters during the period 1986–95 [28]. The goal was to learn from patterns of observations and then make future predictions based on incoming intelligence reports [28]. Given the limited data set and the varying reliability of reports, their goals were modest but useful for predicting future events:

> A statistical analysis based on a simulation of the method showed that the probability of a correct prediction was at best 54%, with an accuracy in predicted position of 5 kilometers and in predicted time of 48 minutes. Prediction rules with a probability and an accuracy such as these should be very useful if they can be approached in practice [28].

Contrast this to the earlier application of selecting a chess move, especially near the end of the game when few pieces remain. In the chess example, the moves are deterministic (a pawn attempting to capture a rook legally will capture it 100% of the time), information is complete (the location of each piece is known with 100% accuracy, 100% of the time), and the information is completely reliable. Further, the evaluation criteria (win by checkmate while avoiding being checkmated first) is clear and constant.

In security applications, machine learning will have to process incomplete information of various (and unknown) accuracy and validity. Its predictions of behavior will not be deterministic, and even the desired outcomes may sometimes be in doubt. The underlying models may be limited or unknown. This is a very different problem than in the chess example, and expectations must be tempered accordingly.

4.2 *Analysis for Nuclear Proliferation*

Specific to the Counterproliferation and Counter-WMD efforts, there may be very limited data sets upon which machine learning may train, at least at the macro level of modelling an entire proliferating states' behavior. The total number of states in history that have attempted nuclear proliferation is in the low double digits, and by the very nature of the action, data about these efforts is often shrouded in secrecy and/or security classification.

In the following sections, constraints on the application of machine learning to general Counter-WMD intelligence and proliferation indicators discussed earlier will be addressed. The feasibility of machine learning will be assessed against the following criteria, based on the discussion in this section:

- Data set size: Is there a data set available large enough to execute machine learning initially? If not, is there a straightforward way of obtaining or creating the needed data for machine learning success?
- Data set similarity: Is the situation to be analyzed with machine learning sufficiently similar to the training data set for the machine to extrapolate trends and develop models to the new problem? As an example, use of the Manhattan Project's uranium enrichment data to feed machine learning for a 21st century proliferator using a commercial nuclear reactor to enrich plutonium is not sufficient.
- Uncertainty: What degree of uncertainty would make the machine learning results useful? Given the other conditions, can this uncertainty be met? Note: this and the data set size criteria are not independent. Large data set sizes can help reduce uncertainty [29].
- Relative Value: Does the machine learning accomplish the task quicker and/or less expensively than hiring analysts at the same financial cost? Does the machine provide insight not possible with human analysts, and does that value exceed its cost? For evaluation purposes, this study assumes the machine learning equipment and technical team costs the same as eight full time human analysts with experience in the relevant area.[2]
- Near Future Potential: Would the utility of machine learning for this indicator greatly increase with modest improvements in supporting technologies that could be expected in the next decade? For instance, if the resolution of satellite imagery was not currently sufficient to identify vehicles in a city, but increasing the resolution by a factor of two would make it possible, machine learning may have great near future potential in that area. Note: this deliberately ignores possible breakthroughs in artificial intelligence, which will be speculated upon in a separate section.

A summary of the initial assessment of machine learning utility and feasibility for the indicators discussed as useful in counterproliferation is shown in Table 1. Each indicator and details of the assessment are further discussed below.

4.2.1 Acquisition of Raw (Unenriched) Special Nuclear Material Precursors

The major precursors of fissile materials are uranium ore, processed uranium, low-enriched uranium, and post-reactor use plutonium [30]. With decades of mining and processing data available, it would not be difficult for a machine learning application

[2]Assuming sufficient information security is available, commercial providers of computational power such as Amazon and Microsoft offer computing power at low enough prices (<$1000/month for most applications) where the cost of associated human operator resources to dwarf machine costs. Since the data sets are not especially massive in the Counter-WMD case, the size of the human analyst section was scoped to be similar to the number of people needed to program, operate, monitor, and interpret machine learning results.

Table 1 Initial assessment of machine learning for counter-WMD applications

	Current			5-10 Years			10-20 Years		
	Data Set	Uncertainty	Value	Data Set	Uncertainty	Value	Data Set	Uncertainty	Value
Raw material acquisition	Fully Capable/Feasible	Fully Capable/Feasible	Mostly Capable/Feasible	Fully Capable/Feasible	Fully Capable/Feasible	Fully Capable/Feasible	Fully Capable/Feasible	Fully Capable/Feasible	Mostly Capable/Feasible
Excess production of low-enrichment U-235	Fully Capable/Feasible	Fully Capable/Feasible	Partially Capable/Feasible	Fully Capable/Feasible	Fully Capable/Feasible	Partially Capable/Feasible	Fully Capable/Feasible	Fully Capable/Feasible	Mostly Capable/Feasible
Excess enrichment levels of U-235	Partially Capable/Feasible	Partially Capable/Feasible	Partially Capable/Feasible	Partially Capable/Feasible	Partially Capable/Feasible	Partially Capable/Feasible	Mostly Capable/Feasible	Mostly Capable/Feasible	Mostly Capable/Feasible
PU-238 short fuel cycle enrichment	Fully Capable/Feasible	Mostly Capable/Feasible	Partially Capable/Feasible	Fully Capable/Feasible	Fully Capable/Feasible	Partially Capable/Feasible	Fully Capable/Feasible	Fully Capable/Feasible	Mostly Capable/Feasible
Power generation excess	Fully Capable/Feasible	Mostly Capable/Feasible	Fully Capable/Feasible	Fully Capable/Feasible	Fully Capable/Feasible	Fully Capable/Feasible	Fully Capable/Feasible	Fully Capable/Feasible	Fully Capable/Feasible
Excavations / hidden facilities	Fully Capable/Feasible	Fully Capable/Feasible	Fully Capable/Feasible	Fully Capable/Feasible	Fully Capable/Feasible	Fully Capable/Feasible	Fully Capable/Feasible	Fully Capable/Feasible	Fully Capable/Feasible
Refusal of international inspections	Partially Capable/Feasible	Partially Capable/Feasible	Partially Capable/Feasible	Partially Capable/Feasible	Partially Capable/Feasible	Partially Capable/Feasible	Partially Capable/Feasible	Partially Capable/Feasible	Partially Capable/Feasible
Nuclear weapons scientists	Mostly Capable/Feasible	Mostly Capable/Feasible	Mostly Capable/Feasible	Fully Capable/Feasible	Fully Capable/Feasible	Fully Capable/Feasible	Fully Capable/Feasible	Fully Capable/Feasible	Fully Capable/Feasible
Nuclear weapon design acquisition	Mostly Capable/Feasible	Mostly Capable/Feasible	Mostly Capable/Feasible	Mostly Capable/Feasible	Mostly Capable/Feasible	Mostly Capable/Feasible	Not Applicable	Not Applicable	Not Applicable
Nuclear Suppliers Group controlled list acquisitions	Fully Capable/Feasible	Mostly Capable/Feasible	Mostly Capable/Feasible	Fully Capable/Feasible	Mostly Capable/Feasible	Mostly Capable/Feasible	Fully Capable/Feasible	Fully Capable/Feasible	Fully Capable/Feasible
Chemical explosives expertise	Mostly Capable/Feasible	Mostly Capable/Feasible	Partially Capable/Feasible	Fully Capable/Feasible	Mostly Capable/Feasible	Mostly Capable/Feasible	Not Applicable	Not Applicable	Not Applicable
Complex electronic expertise	Mostly Capable/Feasible	Mostly Capable/Feasible	Partially Capable/Feasible	Fully Capable/Feasible	Mostly Capable/Feasible	Mostly Capable/Feasible	Not Applicable	Not Applicable	Not Applicable
Delivery system development	Mostly Capable/Feasible	Fully Capable/Feasible	Partially Capable/Feasible	Mostly Capable/Feasible	Fully Capable/Feasible	Mostly Capable/Feasible	Mostly Capable/Feasible	Fully Capable/Feasible	Mostly Capable/Feasible
Nuclear testing	Fully Capable/Feasible	Fully Capable/Feasible	Partially Capable/Feasible	Fully Capable/Feasible	Fully Capable/Feasible	Fully Capable/Feasible	Fully Capable/Feasible	Fully Capable/Feasible	Partially Capable/Feasible
State decision to proliferate	Partially Capable/Feasible	Not Capable/Feasible	Not Capable/Feasible	Partially Capable/Feasible	Partially Capable/Feasible	Partially Capable/Feasible	Fully Capable/Feasible	Mostly Capable/Feasible	Fully Capable/Feasible
State decision to employ	Not Capable/Feasible	Not Capable/Feasible	Not Capable/Feasible	Partially Capable/Feasible	Partially Capable/Feasible	Partially Capable/Feasible	Mostly Capable/Feasible	Mostly Capable/Feasible	Mostly Capable/Feasible

Legend	Colour
Fully Capable/Feasible	green
Mostly Capable/Feasible	yellow
Partially Capable/Feasible	pink
Not Capable/Feasible	black
Not Applicable	grey

to compare these inputs to known expected outputs (nuclear fuel in quantities and enrichment appropriate for civilian reactor and medical isotope generation use.) The machine would be able to detect anomalies from the known data.

However, this is not an especially involved or complex problem, and it could be accomplished in days by a team of human analysts and using spreadsheets. New data is generated slowly and it would be easy to update the assessments. While the capability is there, the relative value of machine learning for this indicator is marginal, and it should not be used. A robust spreadsheet manually updated is less expensive and equally effective.

4.2.2 Enrichment of Uranium in Quantities Exceeding Civilian Nuclear Reactor Needs

This indicator can be obtained not only by international inspections but also by other inputs and outputs of known enrichment facilities. The amount of raw material incoming as well as the amount of "tails" (uranium depleted of ^{235}U) could be observed through typical intelligence means and fed into models of enrichment to confirm or deny the stated enrichment or to help determine enrichment if inspections are refused. In this case, the model is a straightforward mathematical calculation involving logarithms. Machine learning adds no value compared with a human analyst in this case.

4.2.3 Enrichment of Uranium to Levels Beyond the 20% Necessary for Research Reactors or Medical Isotope Production

The application of machine learning to these indicators suffers the same drawbacks as the previous indicator. In addition, there is far less openly available data on enrichment processes exceeding 20% from historical cases [31]. It is also difficult to externally ascertain the efficiency of a given enrichment plant, especially if the proliferating state is manufacturing its own centrifuges or other enrichment equipment, which means the data similarity may suffer and it may not be useful. These factors combine to create unacceptable uncertainty, especially when the significant quantities are so low.

4.2.4 Enrichment of Plutonium Indicated by Artificially Short Fuel Burn Time in Civilian Nuclear Reactors

It is difficult to hide refueling operations in most civilian nuclear reactor designs. The optimal fuel rod burn time for power generation and economic profitability of a CANDU[3] reactor is approximately three years [32]. This greatly exceeds the optimal

[3]CANada Deuterium Uranium reactor.

burn time to create fissile ^{239}Pu, which is far less [33]. Untimely, reduction in power generation from the plant is an easily observed sign of premature refueling, as well as excess amounts of low-enriched fuel being produced and consumed.

It would not be difficult to teach a machine to track these inputs and outputs and determine anomalies. Similar data exists and the underlying burn-up model for commercial reactors is well-understood [34]. However, if a state had a home-brewed reactor, the data would be difficult to analyze if the geometry of the fuel and the neutron flux were unknown. This would flummox machine learning as well as human analysts until more information was known. However, since the model is straightforward and the indicator so clear, a single human analyst would have little difficulty keeping up with worldwide refueling cycles, given the limited number of reactors and infrequent refuelings. Machine learning brings no added utility to this part of the fight.

4.2.5 Excess Power Generation/Distribution in Areas Without a Clear Need (Powering Enrichment Processes Covertly)

Power generation, distribution, and consumption is well-quantified and understood, with large data sets available. The indicator sought is a site of unknown purpose consuming large amounts of energy that could be used to drive enrichment processes. Machine learning is suited to determine when such anomalies exist, and can highlight suspect sites based on known distribution equipment, especially when power lines and transformers exist above ground.

A human analyst team is also well-suited to this type of analysis in a small power network, or a sparsely populated and industrialized one with few generation nodes. However, in a complex urban or industrial environment with multiple power generation sources including off-the-grid capabilities and active power management, it would be possible to hide power consumption in the noise. Machine learning is better suited to handle these large data sets quickly and unusual patterns or anomalies. Human analysts are still needed to help program the model and update the data, but machine learning has great utility here.

It is important to consider that there will always be some uncertainty—machine learning analysis can only indicate suspicious or unknown power usage. Human analysts must fill in the gaps and connect this indicator to others or dedicate collection assets to obtain more information.

The capability of machine learning in this area will grow as power grids become larger and more complex. Also, as sensor technology increases, more accurate power consumption modelling can occur by monitoring waste heat from locations. The laws of thermodynamics dictate that the waste energy must leave the facility somewhere, and without elaborate efforts to mask that signature, it will be more and more detectable and power consumption determined with increasing accuracy.

4.2.6 Evidence of Large Scale Excavations and Other Traditional Intelligence Indicators or Hidden and/or Protected Facilities

This is another area where machine learning can be leveraged to handle large amounts of data quickly. While human analysts likely have the upper hand in evaluating evidence of excavation or hidden facilities on individual images, the machine learning application of imagery analysis is strong enough to triage scores of imagery data quickly into "interesting" and "uninteresting" sets, with the former undergoing further analysis.

There are two factors that will increase the utility of machine learning on this indicator in the future. First, each true positive detected can be fed back into the training data, increasing the accuracy of the machine learning model. Second, increasing resolution and availability of imagery data gives both more and better data to evaluate. Machine learning could make it very difficult to hide facilities, and investments in this area may have significant payoff.

4.2.7 Refusal of International Inspectors or Site Inspection Limitations

While machine learning might be used to analyze patterns of behavior for this indicator, the training data sets available are so small as to make machine learning unworkable. Further, thus far international inspection occur seldom enough that it is easily and better handled by human analysts.

4.2.8 Hiring or Training of Nuclear Weapons Scientists

There are multiple methods for determining the condition of this indicator. Machine learning is well-suited to specific methods of analysis, such as clustering work on cell phone and email metadata, and language recognition if voice conversations can be obtained and possibly decrypted. There are large training data sets available for both, and if the focus is on offshoots of the A. Q. Khan network, known outputs as well. Machine learning could be valuable in tracking a network of proliferation where several key players are known and their communication is somewhat observable. However, there is a lot of uncertainty in this regard. Outside of military institutions, there are few university degrees awarded in "nuclear weapons design," and searching through graduation data and transcripts may not be a useful indicator.[4] However, associations with known proliferators is something that machine learning could possibly track.

The weakness of machine learning in assessing a state hiring or training nuclear weapons scientists is in the sparsity of data and the risk of a dissimilar proliferation

[4] A. Q. Khan earned his Ph.D. in Mechanical Engineering, for instance—too common of a degree to effectively track: [35].

network emerging. However, as more data becomes available over time, adding to data sets can make machine learning more useful in this area in the future.

4.2.9 Seeking or Acquiring Existing Nuclear Weapons Designs

This application could be considered data mining rather than machine learning analysis. Contacts between state officials or representatives (such as universities or labs) and known proliferators create red flags for analysts to assess. The cell phone and email metadata analysis discussed earlier could also be used in this area, but machine learning may be overkill and less valuable than the work of analysts with a filterable database given that the targets are so specific.

There is also the risk that a new design and network may emerge that is too dissimilar to previous proliferators' activities which would render the training data ineffective and potential lead to false negatives. One aspect where machine learning may assist in this application is in discovering unknown relations and nodes in such a proliferation effort.

4.2.10 Acquisition of Specific or Dual-Use Equipment Required for Fabrication and Tracked by the Nuclear Suppliers Group

There should be a large amount of detailed data ripe for machine learning analysis thanks to the Nuclear Suppliers Group. Locating patterns in requests and shipments of key equipment for nuclear proliferation, including indirect and redirected shipments, are well within the capabilities of machine learning. Machine learning would have its best value in looking for unusual patterns in large amounts of seemingly innocuous shipments. Human analysts would easily flag large or suspicious requests and could keep known proliferators under very close supervision (or outright prohibition). However, machine learning may catch unknown proliferators upon which human eyes are not focused.

Machine learning's iterative nature could also provide feedback to the Nuclear Suppliers Group by determining which items are of especially great predictive value of proliferation, and then tracking those items more closely and in greater detail. This may also lead to increasingly valuable predictive value of machine learning in this field over time.

4.2.11 Acquisition or Manufacture of Chemical Explosives in Excess of Conventional Military or Civilian Requirements

The differentiation between conventional high explosives and those used in nuclear weapons is not clear in open source literature. This information is key in determining whether machine learning would have effects similar to that of the nuclear weapons scientist indicator or if it would be ineffective.

4.2.12 Acquisition of Complex Electronics Necessary for Firing Mechanism of an Implosion Device

This indicator is a subset of the weapons design problem, but it is highlighted separately because it requires non-nuclear expertise and special equipment to execute. While there is overlap with the Nuclear Suppliers Group indicator as well as the weapons design indicator, data mining specifically for personnel and equipment for this requirement may yield separate intelligence data that rises above the extraneous noise. However, the author judges that data mining driven by a human analyst is likely better suited than machine learning for this indicator.

4.2.13 Development of Delivery Systems, Including Peaceful Space Programs

There is enough data available on space and missile development programs to attempt a machine learning approach to detecting and evaluating efforts in this area. However, since the events are so infrequent and the signature of a rocket launch or a new aircraft development are so clear, human analysts can easily keep pace with these unique events and analyze them in detail on a case-by-case basis. Machine learning does not add value here.

4.2.14 Evidence of Nuclear Testing (Underground or Undersea Tests)

Nuclear tests in the Non-Proliferation Treaty era are easily detected and gather worldwide attention. There is little difficulty determining that a nuclear test has taken place using currently available means such as seismographs and radionuclide detection. The reason that machine learning could be valuable in imagery analysis of clandestine excavation of material is the large amount of data to be processed and the lack of clear indications of intent. Nuclear testing generally has few events and is very discernable. Machine learning, therefore, would likely not add value here.

4.3 Analysis for Chemical and Biological Proliferation

The indicators related to detection of hidden facilities and tracking knowledge and expertise are most pertinent to the counter chemical and biological WMD fights. The indicators related to fissile material do not apply; nor do the indicators related to complex electronics or testing apply, as chemical and biological weapons can be completed on a much smaller scale and with less evidence than nuclear weapons testing.

Development of chemical or biological weapons delivery systems is an indicator, although a weak one as an adversary is unlikely to use an intercontinental ballistic missile delivery system for chemical or biological weapons, due to expense, risk of increased scrutiny, and risk of nuclear response. Smaller missiles and other delivery systems are more likely to be used with conventional munitions or explosives rather than chemical or biological agents.

5 Potential Uses of Far Future AI in Counter-WMD Applications

The machine learning applications discussed so far are characterized by a computer system manually optimized for the specific indicator analysis required by human experts. It must be provided with properly discretized and verified data to perform a specified analysis. This is the realm of the current day and the near future. Despite decades of effort, the "quantum leap" to a generalized AI system, described below, has not yet occurred.

A generalized AI system, for purposes of this section, is an AI that not only could seek out its own data without specific discretization, formatting, and verification, determine the correct sub-questions to ask, and perform an analysis between data sets in unexpected, unprogrammed, and perhaps even creative directions. This AI can take simply stated objective questions and attempt to answer them without much further guidance. It could also ask for data it that it assesses would be valuable in the analysis. It is unbounded by processing power or data storage in any meaningful way, and thus can handle any amount of useful data, or filter a very large data stream for useful bits. Most distinctly, it can improve its own analysis in a recursive manner as it works—it need not depend on human programmers past the initial setup.

Such an AI could make efforts at far less binary indicators and analyze situations that lack large data sets or precedents. Some possible questions that generalized AI may pursue beyond the current machine learning limitations include the following four issues.

First, has a state's decision maker decided to pursue WMD proliferation? This could be assessed first by determining who the true decision makers are in a state and then using their public and private statements, decisions, records, and surrounding conditions in the state to make an accurate, predictive statement about their desire to proliferate.

Second, how close is a proliferating state to completion or what is the disposition of an established WMD state's arsenal? Generalized AI can bring in a host of factors to determine the level of progress, especially towards nuclear weapons, and identify shortfalls a state may have that could lead to effective interventions. For an established state, knowing where, how many, and what type of WMDs and delivery vehicles are available is the poker game equivalent of knowing what cards an opponent possesses and how deeply buried the card they need is in the deck.

Next, under what conditions would a state use, or not use, WMD? This question is focused on a state's decision making apparatus, and may involve more players than the head of state. Predicting when a state would or would not use nuclear weapons greatly alters the deterrence situation. In terms of game theory, and more loosely poker, it is the equivalent of knowing how much the opponent is willing to bet and when.

Finally, can an international treaty organization be designed and created to severely reduce the likelihood of WMD use or eliminate it altogether? This problem begins with assessing the myriad equities and politics of numerous nuclear or potentially nuclear states. A generalized AI could look for ways to maintain peace through deterrence and/or mutual self-interest, and create a plan that takes incremental steps to accomplish what seems beyond the pale in our current world—states in a long-term, stable peace, which is not dependent upon large WMD arsenals.

6 Recommendations and Ways Forward

Based upon the results of this study, the two indicators that machine learning is best suited to analyze in the Counter-WMD intelligence fight are analyzing power generations and distribution for evidence of enrichment, and imagery analysis to search for evidence of excavation and hiding of facilities.

There is active research on machine learning and the power grid, but it is not clear if there is any focus on analyzing it from an intelligence perspective [36, 37]. If efforts have not begun in adapting machine learning for this type of analysis, the Department of Defense and/or Department of Energy should sponsor university research and development in this area, with the goal of identifying large power consumers and predictive modelling of their consumption. This relatively small investment could create a new, effective intelligence capability.

The state of the art for imagery analysis is surely classified in the US. There has been open source research conducted addressing the use of imagery analysis to detect and determine environmental impacts, including excavation [38]. Machine learning proficiency in imagery analysis has widespread application and utility beyond the sphere of Counter-WMD and development in this area should continue. Intelligence agencies with responsibilities for Counter-WMD should direct some of these research efforts and dollars to detecting hidden facilities if they are not already doing so.

The tracking of key personnel via communications metadata is not a new field, but it is one that can be used to great effect for counterproliferation efforts. It is unknown whether this application is being used currently.

The next steps in this research could involve assessing in detail three areas: power grid analysis, imagery analysis, and communications analysis. Determining the current state of affairs in each area will help drive recommendations for further research. Discussions with experts in the Counter-WMD field for feedback on the analysis thus far could also open new research avenues to pursue. Discussions with machine learning experts, along with their reviews of this research, could also lead to new applications for machine learning and research recommendations.

References

1. US Joint Chiefs of Staff (2013) Joint intelligence. Joint Publication 2-0, 22 Oct 2013. https://www.jcs.mil/Portals/36/Documents/Doctrine/pubs/jp2_0.pdf
2. Nuclear Matters Handbook. Office of the Deputy Secretary of Defense for Nuclear Matters, US Department of Defense, p 105 (2016). https://www.acq.osd.mil/ncbdp/nm//NMHB/index.htm
3. Nuclear Matters Handbook. US Department of Defense, p 175 (2016). https://www.acq.osd.mil/ncbdp/nm//NMHB/index.htm
4. Hoddeson L et al (1993) Critical assembly: a technical history of Los Alamos During the Oppenheimer Years, 1943–1945. Cambridge University Press, New York, p 293
5. Technologies Underlying Weapons of Mass Destruction. US Office of Technology Assessment, OTA-BP-ISC-115, Dec 1993, US Government Printing Office, Washington, D.C.
6. MacCalman M (2016) A.Q. Khan nuclear smuggling network. J Strateg Secur 9(1):104–118
7. Nuclear Matters Handbook. US Department of Defense, p 177 (2016). https://www.acq.osd.mil/ncbdp/nm//NMHB/index.htm
8. About the NSG. Nuclear Suppliers Group. http://www.nsg-online.org/en/. Accessed 11 Dec 2016
9. IAEA Safeguard Glossary. IAEA safeguards glossary, 2001 edition, International Nuclear Verification Series No. 3. International Atomic Energy Agency. https://www-pub.iaea.org/MTCD/publications/PDF/nvs-3-cd/PDF/NVS3_scr.pdf. Accessed 11 Dec 2016
10. Uranium Isotopes. Global Security, 11 Dec 2016. https://www.globalsecurity.org/wmd/intro/u-isotopes.htm
11. Glaser A (2005) About the enrichment limit for research reactor conversion: why 20%? The 27th international meeting on reduced enrichment for research and test reactors (RERTR), 6–10 Nov 2005, Boston, Massachusetts. http://www.princeton.edu/~aglaser/2005aglaser_why20percent.pdf
12. Scott Kemp R (2009) Gas centrifuge theory and development: a review of US programs. Sci Global Secur 17(1):1–19
13. Love O (1973) Electromagnetic separation of isotopes at Oak Ridge. Science 182(4110):343–352
14. Beaton L (1962) The slow-down in nuclear explosive production. New Sci 16(309):141–143
15. Nuclear Weapons Technology Militarily Critical Technologies List (MCTL) Part II: Weapons of Mass Destruction Technologies. Office of the Undersecretary for Acquisition and Technology, US Department of Defense, Feb 1998. https://fas.org/irp/threat/mctl98-2/mctl98-2.pdf
16. IAEA Safeguards Overview. International Atomic Energy Agency. https://www.iaea.org/publications/factsheets/iaea-safeguards-overview. Accessed 3 May 2017
17. Glaser A (2008) Characteristics of the gas centrifuge for uranium enrichment and their relevance for nuclear weapon proliferation. Sci Global Secur 16(1–2):17–18
18. Carson Mark J, Von Hippel F, Lyman E (2009) Explosive properties of reactor-grade plutonium. Sci Global Secur 17(2–3):171–185
19. Nuclear Matters Handbook. Office of the Deputy Assistant Secretary of Defense US Department of Defense for Nuclear Matters, 2016. https://www.acq.osd.mil/ncbdp/nm//NMHB/index.htm
20. Sagan SD (1996–1997) Why do states build nuclear weapons? Three models in search of a bomb. Int Secur 21(3):54–86
21. Control List of Dual-Use Chemical Manufacturing Facilities and Equipment and Related Technology and Software. The Australia Group, May 2017. http://www.australiagroup.net/en/dual_chemicals.html
22. Newborn M (1997) Kasparov versus deep blue: computer chess comes of age. Springer, New York
23. Victor Allis L (1994) Searching for solutions in games and artificial intelligence. Ponsen & Looijen, Wageningen, p 171

24. Alpaydin E (2014) Introduction to machine learning, 3rd edn. The MIT Press, Cambridge, p 9
25. Armstrong H (2015) Machines that learn in the wild. Nesta, 29 June 2015. https://www.nesta.org.uk/report/machines-that-learn-in-the-wild-machine-learning-capabilities-limitations-and-implications/
26. Dunietz J (2016) The fundamental limits of machine learning. Nautilus, 20 Sept 2016. http://nautil.us/blog/the-fundamental-limits-of-machine-learning. Accessed 11 Dec 2016
27. Armstrong H (2015) Machines that learn in the wild, p 7. https://www.nesta.org.uk/report/machines-that-learn-in-the-wild-machine-learning-capabilities-limitations-and-implications/. Accessed 11 Dec 2016
28. Bergsten U et al (1997) Applying data mining and machine learning techniques to submarine intelligence analysis. In: Proceedings of third international conference on knowledge discovery and data mining (KDD 1997), Newport Beach, Aug 1997. AAAI Press, Menlo Park, pp 127–130. https://arxiv.org/ftp/cs/papers/0305/0305022.pdf
29. Guyon I (1997) A scaling law for the validation-set training-set size ratio. AT&T Bell Laboratories
30. Bahtke CG (2012) The attractiveness of materials in advanced nuclear fuel cycles for various proliferation and theft scenarios. Nucl Technol 179(1):5–30
31. Nuclear Weapons Technology Militarily Critical Technologies List (MCTL) Part II: Weapons of Mass Destruction Technologies. US Department of Defense (1998). https://www.hsdl.org/?abstract&did=443125
32. The Nuclear Fuel Cycle. World Nuclear, updated Mar 2017. http://www.world-nuclear.org/information-library/nuclear-fuelcycle/introduction/nuclear-fuel-cycle-overview.aspx
33. Nonproliferation and Arms Control Assessment of Weapons-Usable Fissile Material Storage and Excess Plutonium Disposition Alternatives. Office of Scientific and Technical Information, US Department of Energy, Jan 1997. https://www.osti.gov/servlets/purl/425259
34. Carelli MD, Ingersoll DT (eds) (2014) Handbook of small modular nuclear reactors. Woodhead Publishing, Amsterdam
35. Langewiesche W (2005) The Wrath of Khan. The Atlantic. https://www.theatlantic.com/magazine/archive/2005/11/the-wrath-of-khan/304333/
36. Wu L et al (2011) Evaluating machine learning for improving power grid reliability. In: ICML 2011 workshop on machine learning for global challenges. http://web.mit.edu/rudin/www/docs/WuEtAlNOVA11.pdf. Accessed 11 Dec 2016
37. Gross P et al (2006) Predicting electricity distribution feeder failures using machine learning susceptibility analysis. http://phillong.info/publications/GBAetal06_susc.pdf
38. Petropoulos GP et al (2012) Change detection of surface mining activity and reclamation based on a machine learning approach of multi-temporal Landsat TM imagery. Geocarto Int 28(4):1–20

Directed-Energy Non-Lethal Weapons: An Evaluation of Their Ethical Use and Potential Applications

Seth E. Gordon

Abstract Non-Lethal Weapons (NLW) are not themselves a new or game-changing technology. However, new forms of NLW enable a standoff capability previously only available from traditional or lethal systems. These system architectures rely on directed acoustic or electromagnetic energy to achieve a desired effect in their targets, whether personnel or materiel. As with many emerging and game-changing technologies, political and ethical concerns, alongside technical ones, affect the development of Non-Lethal Weapons. The end result is a debate between NLW advocates who favor utility-based arguments and NLW opponents who point to less quantifiable issues with the use of such systems. This chapter aims to bridge the divide between the two sides of the debate and to illustrate that NLW are not only useful systems but also to explore inherent problems in the three primary arguments against their deployment. NLW are a greater ethical good than either lethal (or permanently-harming) force or the lack of action that results in further losses. Politically, the treaties that would govern NLW deployment and use either do not address these systems directly or indirectly encourage their use by advocating for minimizing collateral damage. Lastly, concerns about the "slippery slope" of NLW technology development is a logical fallacy based upon the erroneous concept of "cerebralcentrism" and the assumption that NLW provide an oppression capability not currently possible with traditional lethal force.

1 Introduction

> The United States can feel secure that it has the firepower to defeat any invading army. The question is, can we stop a conflict without excessive collateral damage and without exposing our forces to excessive risk?—D. V. Giri [1]

S. E. Gordon (✉)
Georgia Institute of Technology, Atlanta 30332, GA, USA
e-mail: seth_gordon1@mac.com

M. E. Kosal (ed.), *Disruptive and Game Changing Technologies in Modern Warfare*, Advanced Sciences and Technologies for Security Applications,
https://doi.org/10.1007/978-3-030-28342-1_6

As compared to many emerging and game-changing technologies, political and ethical concerns, rather than technical concerns, have hindered the development of Non-Lethal Weapons (NLW). More specifically, opposition to NLW centers on in-development "directed energy" systems that rely on either electromagnetic (EM) or acoustic vectors to achieve a desired effect. Major non-governmental organizations (NGOs) and a variety of international organizations and treaties impede research into and involving NLW. Example opposing organizations and their justifications include:

- The American Red Cross asserts that NLW inflict an unethical level of pain on their targets [2]
- Human Rights Watch asserts that NLW use in a "riot-control" context can stifle the expression of Basic Human Rights [2]
- The American Civil Liberties Union has similar reasons to the above, but also has claimed that "the road to hell is paved with [...]" good intentioned Non-Lethal Weapons [3]
- The U.N. Committee Against Torture identified Tasers and Taser-like NLWs as almost a form of torture and advocated against their continued use [4]
- The Federation of American Scientists suggests that NLW not only cause unethical levels of pain but also that the systems have limited utility [3]

The proponents of NLW focus on the utility of these systems but do not directly address the opposition arguments that tend to instead focus on the ethical, moral, and legal implications. In discussing and advocating for the use of NLW, it is necessary to not only review their utility but to also to adequately defend their use against ethical and legal counterarguments. This chapter aims to firstly discuss the utility of NLW before articulating their legality and ethical nature as compared to traditional conventional systems. After all, a new weapon does not exist in a vacuum, but rather must be compared against the systems it is either replacing or supplementing.

2 Example Non-Lethal Weapon Architectures

> The Department of Defense defines Non-Lethal Weapons as weapons, devices, and munitions that are explicitly designed and primarily employed to incapacitate targeted personnel or materiel immediately, while minimizing fatalities, permanent injury to personnel, and undesired damage to property in the target area or environment.—United States Department of Defense Directive 3000.03E [5]

To enable a discussion about the utility, applications, and ethics of NLW, it is necessary to describe possible system architectures. Furthermore, one must compare NLW to conventional munitions to enable proper evaluation and discussion of these weapons. For purposes of the comparison, this chapter divides conventional weapons into three different forms. The first are unguided, direct damage systems. These include everything from bullets, to artillery shells, to directed-energy weapons. These systems, while perhaps guided at launch (or firing) are ultimately "dumb" weapons whose capabilities after firing are entirely dependent on physics. The second group includes guided, but unpowered weapons, such as guided bombs, guided artillery shells, or ballistic missiles. While control is possible during flight to target, these munitions do not possess independent propulsion and are reliant on potential and kinetic energy to reach their targets. Lastly, there are guided and powered weapons which conventionally include various types of missiles. These systems not only have guidance for the duration of their travel to target, but they also possess the propulsion system(s) necessary to adjust or maintain their flight.

On the NLW front, there are two axes of comparison and breakdown. The first axis is the system application and whether it is intended to be counter-materiel versus counter-personnel. As the titles suggest, this division is a function of whether a system is intended to engage equipment or people, respectively. The second axis is the form of this interaction. For the requisite analysis, three forms of NLW architecture are detailed below.

The first form is passive NLW, which includes caltrops, spike strips, counter-traction technologies, and the like. These systems are all similar in that once the NLW has been deployed, it requires no active control to engage or interact with its target. There is no chemical interaction between materials or input of energy necessary as the interactions are instead more traditionally physical. Example architectures include Anti-Traction Technologies (A-TT) or "Super Adhesives" which dramatically reduce and increase friction, respectively. Such systems can be applied to road surfaces or vulnerable components to introduce hazards to enemy equipment operation [1].

Furthermore, Combustion Alteration Technologies (CAT) can prevent traditional combustion processes and therefore stall engines for as long as the compound remains present to inhibit standard operation [1]. Lastly, foam and "entanglement" technologies can prevent physical movement of either materiel or personnel, but they are physical-restriction based rather than physiological or chemical [1]. All of these NLW are notable for generating effects following active application by the operator. They are almost always inherently reversible; though there is risk of damage should opponents attempt to operate equipment in spite of NLW use (e.g., car crashes due to A-TTs). Passive NLW are also usually counter-materiel in purpose with the intention of making asset operation dangerous or impossible rather than destroying the equipment directly.

The second form is active conventional NLW. These compounds either induce reactions (sedation, irritation, nausea, etc.) in humans or cause harm on a molecular level to equipment and materials (liquid metal embrittlement, etc.). Similar to passive NLW, an active input from the original user is not necessary, though the actual non-lethal effects are the result of active interactions between the applied substances and the targets, usually with deeper effects than those of passive NLW. These active systems can include applications from passive NLW such as putting A-TTs into a landmine-like deployment system [6]. Active conventional NLW can also affect both personnel and materiel. Liquid Metal Embrittlement (LME) induces a chemical reaction that dramatically weakens the components of a targeted system, making it either dangerous of impossible to operate. Supercaustics are similar but instead deteriorate systems and subsystems more directly [1]. On the counter-personnel front, malodorants can discourage human activities in a given region while calmative agents can sedate opponents [1]. Active conventional NLW also include more traditional "riot control" systems such as tear gas or Tasers. These systems all require active user deployment or ongoing interaction and usually cause more explicit chemical, physical, or physiological effects than the passive NLW discussed previously.

The third form is directed energy NLW and the primary subject of this chapter. These systems use either specific electromagnetic or acoustic means to transmit energy to the target and cause a desired effect. These systems range from strobes to induce confusion to an electromagnetic weapon intended to disable vulnerable electrical equipment. Furthermore, this category of NLW is broad and ranges from area-effect systems such as the Active Denial System (ADS) to more targeted "pulsed energy projectiles," which are directed against single targets [7]. On the counter-personnel front, NLW architectures are usually built around acoustic or electromagnetic energy-transfer means with the intention of causing intense discomfort in opposing personnel.

Acoustic systems aim for disorientation or nausea while electromagnetic systems usually aim for inflicting pain with no actual physical harm [1]. However, other electromagnetic systems such as the Low Energy Laser (LEL), Isotropic Radiators, or Visual Stimulation and Illusion (VSI) instead can cause temporary blindness or disorientation in a target due to the bright flash or strobing (Bucha effect) [1]. These NLW can be effectively applied in a defensive position in order to discourage opponent attack. EM counter-materiel NLW are of interest because a targeted electromagnetic signal can damage or disable vulnerable electrical systems [1]. While the electromagnetic pulse (EMP) is a widely cited form of counter-material NLW, it is at present a difficult effect to produce and control short of detonating a nuclear device.

Lastly, it is necessary to distinguish between NLW and "cyber" weapons or attacks. Hacking or compromising electrical and computer systems can achieve non-lethal results. However, any counter-material cyber-attacks utilize the hacked systems themselves to inflict the damage. For example, the Stuxnet computer virus directed against the Iranian nuclear program would instruct centrifuges to spin beyond their capabilities, leading to (in theory and intention, but unconfirmed) structural failure [8]. Conversely, NLW inflict the damage due to the actions of the "generating" weapon rather than the "target" equipment. In more qualitative terms, a Non-Lethal Weapon is a system that can be deployed in the field against an opponent while a "cyber" weapon is not necessarily a piece of hardware on the battlefield.

The remainder of this analysis focuses on the active directed-energy NLW, though other technologies will be discussed where appropriate. It is these electromagnetic and acoustic systems that have generated the most ethical backlash from parties such as the International Committee on the Red Cross and the United Nations. Furthermore, these forms of NLW are the ones that are most promising in a variety of scenarios due to both their capabilities ("firepower" one could say) and their ability to project effects at range and in a reusable manner. Passive NLW are less capable in an offensive capacity while active conventional NLW are usually not reusable. In comparison, active directed-energy NLW, given a sufficient power supply and system integrity, can operate indefinitely.

3 The Firepower Dilemma

> I suppose it is tempting, if the only tool you have is a hammer, to treat everything as if it were a nail.—Abraham Maslow [9]

Given a sufficient understanding of the technologies and systems described as NLW, the next step is to consider their utility and the other forms of weaponry available to would-be users.

In the late 20th and early 21st centuries, the United States faces a problem with matching technologies and assets to the capabilities demanded of them. Recent conflicts have illustrated the two-pronged weakness in the focus on raw military firepower. First, exhibitions of such power can exacerbate the "Security Dilemma" as regional powers observe and develop countermeasures to the capabilities on display by the United States. Such a scenario is epitomized by the reactions of Russia and China to the 1991 Persian Gulf War [3]. Second, overwhelming firepower, especially when delivered by aircraft or standoff munitions alone, has proven less than ideal when checking the actions undertaken by non-state actors such as the Islamic State (ISIS). Such approaches are effective when applied in conjunction with ground forces that can hold territory, but otherwise can be counterproductive as "collateral damage" can bolster support for opposing forces disproportionate to the damage actually inflicted by a given strike [10].

In short, and especially in the 21st Century, further increases in conventional firepower do not promise the same gains experienced in prior decades. Instead, developments in novel technologies and investments in Special Operations Forces (SOF) have enabled a degree of power projection against opponents independent of form or stature. However, some principles remain relevant as the modern battlefield shifts in regard to both participants and form. While many prior and current research endeavors focus on maximizing those properties of conventional munitions (e.g., guided artillery shells like Northrop Grumman's XM395 Precision Mortar), ongoing research seeks the same principles with different means [11]. In fact, NLW fit perfectly into the discourse as such systems work toward the desired thresholds in accuracy, precision, and controllability, albeit in a distinctly unconventional manner.

The accuracy of a weapon is reflected in its ability to hit the desired target. For conventional munitions, this property is reflected by the complex aiming, training, and guidance systems used when engaging opponents. Significant technology investment has led to lower levels of collateral damage, especially when compared to the "carpet-bombing" or saturation campaigns seen during the Second World War and the Vietnam War [12]. With regard to NLW, accuracy is usually as good if not superior to conventional weapons. By basic logic, if the NLW is deployed via a pseudo-conventional system (e.g., tear gas grenade), then the system is on par with its lethal counterpart (e.g., grenade launcher). However, new directed-energy NLW architectures have dramatically improved accuracies. Electromagnetic and acoustic energy transfers suffer far less from, or are indeed commonly immune to, environmental effects (e.g., wind) than physical rounds and also can be tuned to affect very specific regions of space. As a result, some current and most future NLWs promise a level of accuracy that, at present, exceeds what conventional munitions can achieve.

Precision is a weapon property describing how consistently a weapon performs. In this regard, precision includes more traditional metrics such as ballistic dispersion but also includes elements such as level of damage or lethality [13]. Furthermore, lethal weapons aim to inflict lethal harm. In conventional munitions, this kind of impact-effect precision is usually taken for granted, as kinetic effects tend to affect the human body consistently regardless of the individual in question.

Conversely, NLW can, depending on the architecture, have varying levels of impact based on the target's physiology or pain tolerance threshold. Due to the desired non-lethal effects of NLW, there is a target range of effect that needs to be achieved in a more precise manner in order to avoid either a lack of effect or an unacceptable risk of lethality. As a result, it is difficult to develop the dosage or weapon setting for the desired effect. Russia discovered this exact problem when they attempted to use a chemical sedative prior to a hostage rescue operation in a Moscow theater in 2002 [14].

Lastly, precision can also include the literal physical precision in that only the desired target is affected. For NLW delivered by pseudo-conventional means, the precision is as good as traditional weapons. With regard to energy-based systems, physical precision is usually high, though most counter-personnel NLW suffer from the aforementioned inconsistent performance against inherently variable human targets.

The final weapon property of interest is controllability. Controllability can represent a variety of properties but can be effectively interpreted as the ability to achieve a given result through modulated user input. Guided munitions are inherently highly controllable as they can be directed to their targets in a way that ensures both high accuracy and high precision [13]. However, such controllability does not account for a potential second axis of the problem: that of controlling the (destructive) power of the munitions. While some weapon systems can equip different munitions to achieve a measure of controllability along this axis, NLW promise a level of control not previously experienced by military forces.

While active conventional and passive NLW have the same controllability problems of conventional munitions, directed-energy systems are known for their ability to enable so called "graduated lethality" [15]. In short, a "graduated lethality" capability allows for a single weapon system to induce a range of warning, non-lethal, and lethal effects with the "push of a button." In comparison, a traditional and inherently lethal weapon can be aimed in an attempt to generate non-lethal effect, but there are no guarantees.

4 Case Study Evaluations and Comparisons

> The general who wins the battle makes many calculations in his temple before the battle is fought. The general who loses makes but few calculations beforehand.—Sun Tzu [16]

The central discussion on the utility of NLW hinges on describing and discussing their applicability to the scenarios potentially facing military forces in the 21st Century. The next few sections will walk through the utility-based arguments around NLW. Lastly, there is an additional utility-based argument for NLW development as a means to understand their capabilities and thus have sufficient countermeasures prepared in case of their use by opponents.

As part of that effort, six scenarios of interest are outlined that are further subdivided into three broad forms. The six scenarios (with associated modern archetypes) and the three higher-level groupings are:

(1) Nation State and Pseudo-Nation-State Engagements

- (a) Peer Competitor—China
- (b) Hybrid Warfare—Conflict in Eastern Ukraine

(2) Non-State and/or Insurgent Situations

- (a) Desert Insurgency—Islamic State and Syrian Civil War
- (b) Non-State Actor—Organized Criminal or Cartel Elements in the Global South

(3) Urban Engagements

- (a) Mega Cities—Conflict or Engagement in a Metropolis With a Population Exceeding 10 Million People
- (b) Feral Cities—Urban Ungoverned Spaces.

In this breakdown, there are clear themes linking each pair of scenarios. Scenarios 1a and 1b are unified in the issues concerning more traditional and mechanized international military engagements against nation-states or against proxies that operate with "all but explicit" sanction of nation-state sponsors. Scenarios 2a and 2b are missions against non-state actors that are entrenched, dispersed, and usually motivated by more local concerns. Lastly, scenarios 3a and 3b concern urban settings with variable levels of development or level of control. These scenarios are not mutually exclusive, though each is best referenced and discussed in conjunction with their associated archetype. The next sections will review the case studies in further detail.

It is useful to compare the estimated utility and applications of various forms of directed-energy Non-Lethal Weapons against the technology's maturity (commonly known as the Technology Readiness Level or TRL). Generally speaking, most of the directed-energy NLW have high utility in a defensive context, both due to their power supply requirements (easier to achieve at a facility than in the field) and their deterrent capability. Counter-personnel NLW work most effectively in situations where opposing forces are aggressive while counter-materiel NLW are most effective against opponents who rely on technological systems that are vulnerable to such a weapon.

A major question that should be addressed when considering NLW use is "what are the Rules of Engagement (RoE)?" Given the specific guidelines and rules provided to combatants regarding conventional weapons, there is a clear and present need to adapt said rules to the use of NLW. Relevant concerns include:

- Where do individual NLW systems lie along the "use of force continuum"?
- What warnings, if any, are necessary before NLW use on a target?
- When is it appropriate to train an NLW (as compared to a conventional weapon system) on a potential threat? (relevant because NLW are frequently non-threatening in physical appearance)
- What settings for the NLW are appropriate to use in different scenarios, assuming graduated lethality is a capability?

Due to the more limited objectives of the associated research, these questions are not answered in this chapter. However, these concerns are relevant to the actual deployment of NLW once the technology has overcome the ethical and political hurdles detailed in the following sections.

4.1 Utility Against a Peer Competitor

> As we have repeated at every Ministerial and Summit since Russia launched its campaign of aggression against Ukraine, NATO Allies stand firm in our support of Ukraine's sovereignty and territorial integrity—Former United States Secretary of State Rex Tillerson [17]

Non-Lethal Weapons are of immense utility against state actors and in "hybrid warfare" scenarios as epitomized by the conflict in Eastern Ukraine. Notably, these paired scenarios are dominated by the presence and use of mechanized or otherwise modernized forces. As a result, electromagnetic counter-materiel systems are especially useful as they are effective deterrents to military actions. A hypothetical Chinese invasion of Taiwan would be stalled if an EMP were to disable all the assets necessary for invasion. Similarly, area-effect counter-materiel NLW can prevent the movement of enemy goods or advanced equipment, such as the missile system used to bring down Malaysian Airlines Flight 17 [18].

Notably, these scenarios imply more traditional "open-field" combat as compared to the other four scenarios. In a recent historical context, these would include the tank battles across Iraq during the 1991 Persian Gulf War. On a more historical front, these imply blue-water naval engagements rather than littoral or small-boat combat seen during the War on Terror. In general, the included scenarios imply a degree of parity between the forces involved. Though the United States would hope to maintain a competitive edge, it remains more difficult to conduct operations against a nation with, for example, a functioning Integrated Air Defense System (IADS) than against one without.

In a present-day context, the conflict in Eastern Ukraine and military posturing in East Asia both illustrate that such force-on-force engagements and interactions are possible. The conflict in Eastern Ukraine does involve urban elements, though there is also open-terrain and long-range vehicle combat and supply lines that are vulnerable or relevant to the conflict. Military activities in East Asia range from possible land-based force-on-force combat in the Korean Peninsula, to another "Taiwan Crisis," to the naval and airstrip buildup in the South China Sea. Again, the more mechanized nature of these engagements increases the potential utility counter-materiel Non-Lethal Weapons. Though military forces strive for resiliency, any modern force is more vulnerable to advanced EM NLW than a guerilla force with AK-47s.

Despite the more mechanized nature of these hypothetical engagements, counter-personnel NLW remain useful, though to a lesser degree than in the caves or urban environments detailed in the other scenarios. However, directed-energy counter-personnel NLW are useful in a defensive position to deter enemy actions. As in the case of a hypothetical nation-state invasion or the attempts to seize territory, deterrence at the individual combatant level could help slow opponent aggression to enable sufficient buildup of conventional lethal systems.

4.2 *Utility in Rural or Non-urban Insurgent Situations*

> If in order to kill the enemy you have to kill an innocent, don't take the shot. Don't create more enemies than you take out by some immoral act.—Former United States Secretary of Defense James Mattis (General, United States Marine Corps, Retired) [19]

The areas of operation for these scenarios are often relatively undeveloped, which fortunately minimizes collateral damage from the use of area-effect NLW. While desert or jungle insurgencies and organized crime often possess decent technological sophistication, they are usually not as vulnerable to counter-material energy-based NLW as more conventional forces. Thus the geography makes NLW an attractive tool, assuming they had targets to engage.

It is remiss, however, to ignore the disparate terrain and natural environments inherent in these scenarios. The dry and scorching desert is taxing on equipment in a way that dense and humid jungle is not. However, both engagement areas are similar in both the form of expected combat and the inclusion of the weather and environment as complicating factors.

Regardless of the counter-materiel NLW possibilities, counter-personnel NLW are of immense utility to help resolve hostage situations, bypass human shields, or capture high value targets. In the caves or bunkers used by opponents, counter-personnel NLW are the perfect tool to "flush" out or incapacitate targets.

4.3 Utility in an Urban Areas

> Every room is a new battle. … Avoid cities if you can. If you can't, avoid enemy areas. If you can't do that, avoid entering buildings.—Unidentified Israeli Military Officer [20]

Non-Lethal Weapons are useful in an urban engagement because of the risk of and use of civilians as human shields. Urban situations also have large civilian populations in the area of engagement. High structural density or tall buildings both present difficulties as allied forces seek to progressively secure and pacify large and complex regions. Booby traps, mines, and ambushes are all prevalent in urban areas where the element of surprise and asymmetrical approaches combine to make the situation dangerous for military forces.

In dense urban environments, the United States' traditional military advantages in mobility and firepower are challenged. Helicopters are vulnerable to ground fire and have difficultly maneuvering between buildings or in dense urban canyons. Armored vehicles are vulnerable to three-dimensional engagement when opponents effectively garrison a variety of structures. Furthermore, vehicles might not be able to sufficiently maneuver along alleyways and other narrow passages. Lastly, the raw firepower of artillery, airstrikes, vehicles, etc. are effected by the desire to minimize collateral damage, risk of structural collapse, or risk of damage, which will further hamper the aforementioned minimal vehicle mobility.

These weaknesses can hopefully be offset through the application of NLW. Though structures offer a degree of shielding from both EM and acoustic NLW, the revolutionary change is that NLW can affect target personnel and materiel without directly compromising the structures housing them. While it is perhaps convenient to simply level a building housing critical enemy forces and infrastructure, it is more utilitarian to remove the opponent while leaving the structure intact for future civilian and peaceful use. However, Mega Cities present further difficulties with the size and height of the affected structures or by the ad hoc nature of shantytowns or other improvised spaces. In both cases, systematic clearing is both necessary and near impossible with conventional approaches. EM and acoustic NLW can hasten the process by implicitly protecting allied forces either by incapacitating opponent personnel or disabling opponent weaponry before damage can be dealt.

Furthermore, NLW enable new battlefield options with lower permanent casualties than any prior weapon system, which is especially relevant in high-density urban environments [6]. As per a "Blackhawk Down" insurgent scenario, mission success may be the rescue of hostages. In a similar situation, the USMC accomplished a successful evacuation of United Nations' personnel from Mogadishu through clear communication and strategic deployment of NLW to discourage civilian interference in an otherwise peaceful action [21]. In short, while current concept of operations (CONOPS) implies that violent force will be used, it is often not actually required by the overall mission objective.

5 Technological Development as a Countermeasure

> Physics and chemistry work the same for everybody and there are very few technical secrets that last very long.—John Alexander (Colonel, United States Army, Retired) [3]

There are two reasons to develop a technology. The first is the most common and is essentially researching a technology in order to use it directly. However, the second reason is more nuanced and is that in developing technology, you can also learn how to counter its use. In the case of Non-Lethal Weapons that second reason is immensely relevant. While possessing NLW is a critical new capability for our men and women in uniform, being able to counter and mitigate their use is important on a defensive level and in terms of protection and survivability.

In past eras, "codes of war" have been pursued to attempt to ensure appropriate and ethical treatment of opponents and non-combatants. In the modern era, the Geneva Conventions and other international laws of armed conflict aim to ensure that prisoners of war, civilians, aid workers, and the like are protected and that the consequences are grave if that protection is violated [22]. However, the dominant conflicts of the recent decades have involved non-state actors with moral and ethical codes all but incompatible with the international order. As such, there is little to no guarantee of the safety of prisoners of war, as most dramatically seen in the beheadings and immolations of journalists and captured servicemen at the hands of the Islamic State.

As such, the capture of an American military service member by such a group could be a severe blow to the United States and a propaganda coup to the group who did the capturing. To enable such a capture, NLW could be employed against the military advisors, Special Operations Forces, and other by non-state actors. By their very nature, NLW can render the most capable warfighter impotent on the battlefield, negating the training and technology currently at the disposal of the American or allied soldier. Without continued research into NLW, it is all but impossible to study or test potential countermeasures. Until countermeasures are developed, soldiers and civilians remain at risk of such weapon systems.

The historical development of protective aviation technology can be understood as a parallel to this argument in favor of continued research into Non-Lethal Weapons. As great as the US defense industry is today, it was in many ways behind its allies and rivals on the eve of World War II. Though some aircraft would prove to be incredibly capable, budgetary issues prevented their smooth or swift introduction to front-line units [23]. Additionally, lack of investment in aircraft technologies cycled into a lack of investment in counter-aircraft or anti-aircraft (AA) technologies. In fact, the outdated AA systems made Pearl Harbor all the more vulnerable to the concentrated air attack of December 7, 1941 when the weaponry around the harbor proved minimally capable of effective defense [23]. In short, research into a novel technology is required for both effective developments of the field and in investigating the necessary means to counter the technology if used by opponents. It is a mistake that the United States can ill-afford given the current global environment.

As devastating as the use of NLW by non-state groups would be, however, there is absolutely no indication that even state actors have access to or the capability of developing field-capable and area-effect Non-Lethal Weapons. In a similar manner though, there are still ample preparations against WMD attacks, even though non-state actors and terrorist groups have similar levels of non-access to the requisite technologies [24]. In this same manner, research can and should continue for NLW, at a minimum to enable the development of countermeasures. This view is reflected in the language of the Chemical Weapons Convention (CWC) wherein signatories are allowed to maintain or manufacture otherwise banned materials to enable the development of countermeasures [25]. Regardless, a near- or peer-competitor is likely to possess NLWs.

6 Treaty Obligations Related to Non-Lethal Weapons

> Even though Article IV of the Constitution says that treaties are the 'supreme law of the land,' in most instances they're not even law.—Supreme Court Justice Sonia Sotomayor [26]

The prior sections have presented the technology behind NLW and the utility-focused, pro-NLW, arguments around their development. However, as mentioned in the introduction, NLW development has been severely impacted by non-technological elements. Treaties, international agreements, the laws of armed conflict, the Geneva Conventions, and public opinion are the centerpieces of most arguments against further NLW development. However, this work asserts that NLW are ultimately more ethical than, as well as equally lawful to, conventional munitions, so each counter-NLW argument will be addressed and countered.

Several treaties explicitly affect the conduct of warfare. On the explicit front, the Geneva Conventions, the Chemical Weapons Convention (CWC), and the Protocol on Blinding Laser Weapons (PBLW) are the most applicable to Non-Lethal Weapons. However, these agreements also have implicit effects and restrictions given the forms and actions of warfare they support or reject [27]. Most notably, the agreements address any "incapacitating" chemical weapons (but not explicitly the new energy-based ones under development) and any actions that impact civilians (area-effect NLW will affect anyone in their target region).

The Geneva Conventions cover conduct in wartime in the broadest sense [22]. There are technically four Geneva Conventions that ensure the safety of wounded or medical personnel on land (First Geneva Convention), those peoples at sea (Second Geneva Convention), prisoners of war (Third Geneva Convention), and civilian populations (Fourth Geneva Convention). These four primary conventions, while extensively governing the conduct of military forces, do not expressly or even implicitly affect the use of the Non-Lethal Weapons. Instead, that effect comes from the so-called Article 3 of all four conventions, which requires the humane treatment of all persons without distinction and specifically bans murder, mutilation, torture, cruel treatment, hostage taking, and unfair "trials" [28]. It is the various Article 3 elements that complicate the deployment of NLW as groups such as the United Nations Committee Against Torture have deemed Taser-like incapacitation devices as "torture" [3]. More generally, the development of "pain-inducing" NLW like the Active Denial System could be seen to run counter to this element of the Geneva Conventions as pain without intent to kill can be considered a form of "cruel treatment."

The Chemical Weapons Convention is of interest when discussing NLW because it explicitly governs the use of incapacitating "Riot Control Agents" (RCAs) in wartime [25]. Entering into force in 1997, the CWC is an extensive document, but of most relevance to this analysis is Article 1, wherein the treaty signatories agree "not to use riot control agents as a method of warfare." Article 2 defines RCAs as:

> 'Riot Control Agent' means: Any chemical not listed in a Schedule, which can produce rapidly in humans sensory irritation or disabling physical effects which disappear within a short time following termination of exposure [29].

In short, an RCA seems to be almost a textbook definition of a Non-Lethal Weapon. Consequently, this complicates the use of any chemical-based NLW, even though signatories can employ them in domestic law enforcement situations. Regardless, and similar to Article 3 of the Geneva Conventions before it, the CWC complicates the deployment of Non-Lethal Weapons by discouraging the *effects* of such systems. As compared to historical arms control treaties, which govern the use of *specific weapon systems* (nuclear devices, specific chemical compounds, specific warship types), the RCA component of the CWC focuses on the effects rather than the source of the incapacitation. Consequently, the development of a non-chemical NLW with the same set of effects can be seen as an improper violation of the spirit of the CWC.

The third international agreement with significant influence on the deployment and development of NLW is the Protocol on Blinding Laser Weapons of 1980, which specifically prohibits the use of laser light to induce permanent blindness in an individual, though exceptions are made for potential secondary effects of laser weapons (e.g., use against an opponent optical system which both damages hardware and may blind the operator) [30]. While this international agreement is narrow in its focus, it has a similar complicating effect on NLW as the Chemical Weapons Convention. Much like the CWC, the official focus is on a specific architecture of NLW. This similarity is further compounded by the risks associated with new weapons violating the spirit of the law. For example, the in-development Distributed Sound and Light Array employs lasers, dazzlers, loudspeakers, and other systems to discourage further advance upon a military-occupied position [31]. While such a system has little to no risk of direct fatal harm to is targets, this NLW does inherently risk vision or hearing-related injuries simply through the form of deterrence utilized.

These three treaty obligations cover a wide range of possible NLW effects and architectures. However, their existence is emblematic of the novelty of Non-Lethal Weapons given that each agreement only officially addresses singular weapon architectures. The intention of these laws is to limit the range of effects engagements can have on both civilian and combatant populations. Regardless, these agreements implicitly limit the use of NLW and provide a backbone upon which future restrictions could be based.

In response to these treaty restrictions, it is necessary to note that only the CWC specifically addresses a Non-Lethal Weapon type that is both in service and that has sufficient data to verify effects. The other treaties do not explicitly cover the energy and sound-based NLW in development, especially given that the ADS and other systems are designed to have reversible effects, unlike those prohibited by the Protocol on Blinding Laser Weapons. Lastly, the Geneva Conventions and other laws of armed conflict are rooted in the assumption that any engagement between a combatant and a civilian creates the risk of permanent harm. NLW specifically short-circuits that assumption by shifting the magnitude of the interaction from permanent, violent harm to (possible) reversible, temporary effects.

7 The "Slippery Slope" of NLW

> Cerebralcentrism [... is the predominantly] American trait in which we believe ourselves to be the only smart people in the world [... meaning] if we don't develop these weapons, no one else can.—John Alexander (Colonel, United States Army, Retired) [3]

As with any new technology, detractors have concern regarding potential future and illiberal applications. In the case of Non-Lethal Weapons, opposition to such systems hinges on the potential to enable human rights abuses or more generally, the unethical control of a populace. Example nations where such heavy-handed tactics are possible include Venezuela, North Korea, Iran, and more generally nations where fear and force are necessary to maintain state authority [3]. The Red Cross highlights possible use by such authoritarian regimes when opposing NLW and has campaigned against the further development of these weapons [32].

The foundation of a "slippery slope" argument is a chain of seemingly logical situations that lead from the current state of affairs to an undesirable (or less often a desirable) outcome [33]. However, the fundamental fallacy of this argument is that there is little to no evidence that each element in the chain will come to pass. There is also usually scant evidence that such a chain is actually possible in the first place. While such an approach has the appropriate hallmarks of a well-reasoned argument, a "slippery slope" case is impossible to confirm and so should not form the basis of decision-making.

In the case of Non-Lethal Weapons, it is improper to assume manifestly that their continued development will lead to less free states and peoples. While certainly NLW have historically and will continue to be used as riot-control measures, that does not independently imply a nation-state or people will use them for oppressive purposes. In fact, given the proliferation of arms and conventional military technology, such oppression already exists under Islamic State, North Korea, and other authoritarian ruling groups around the world [3]. Restricting the capabilities of the United States and allies because of the potential use of technology by rogue parties is antithetical to the eventual containment of these groups as it would simply enable such organizations to proceed with their own programs unchecked.

In his essay on the ethics of Non-Lethal Weapons, John Alexander notes this fallacy. He invokes the fallacy of "Cerebralcentrism," which he describes as a phenomenon where a people believe themselves "the only smart people on the planet" [3]. Furthermore, he asserts this belief seems to be more concentrated in the United States than in other nations, possibly as a result of the memory of the Manhattan Project forever influencing scientific advancement. Regardless, it is erroneous to believe that halting research in a particular field in a particular nation will prevent the maturation of the relevant technologies. Most notably, bans on human genetic engineering in the West has not halted the progression of research and currently China is at the forefront of embryo genetics [34].

Furthermore, cutting-edge technologies almost always have multiple capabilities, and Non-Lethal Weapons are no different. While export controls and other agreements already restrict the trade in similar dual-use technologies, they do not necessarily impact the research on these technologies in the first place within a single nation. This can further the security dilemma as progress on NLW work will remain in a classified or otherwise restricted realm, leading to an ambiguous capability in the eyes of would be rivals and opponents. In a similar manner, NLW can and should be regulated, thus addressing one form of the opposition to NLW. International agreements can restrict the use of or research in a technology, but the threat alone of such agreements should not restrict the development of the affected technologies.

Artificially restricting technology before understanding its demonstrated costs and benefits is shortsighted. As dangerous as nuclear weapons are, the initial weapons research has contributed to the rise of peaceful atomic technologies including electricity generation, medical diagnostics, and spacecraft systems. At this stage, it is difficult to tell whether NLW could usher in such peaceful applications. However, the supporting technologies in energy generation and storage would have undeniable benefit given the ubiquity of electricity in modern society. Regardless, NLW should remain in development because there are minimal signs such systems are worse than current military weapon architectures.

> In short, only the logical fallacy that underlies the "slippery slope" argument indicates that NLW will cause more harm than good to our world. Thus, another roadblock to the development and deployment of NLW has been moved aside.

8 Ethics of Non-Lethal Weapons

> Some members of the military hold the perspective that the concept of nonlethal is antithetical to war itself. It is likely that a portion of the American public hold similar perspectives.—D. V. Giri [1]

A grand irony is at play when discussing the ethics of warfare. There is a universal disgust at needless violence and killing, though the definition of *needless* will vary culture to culture. Conversely, there is a similarly universal adoration of the noble warrior, though again, the definition of *noble* and *warrior* can and will vary culture to culture [35]. This *noble* warrior is necessarily distinguished from *murderers*, *killers*, and such revolting individuals who also kill. Given these apparent contradictions, is there a possible middle ground where the *noble warrior* can accomplish his mission without killing? Non-Lethal Weapons promise to provide the necessary tools.

Exactly as their name suggests, NLW are a means of engaging a target without killing the human operator or combatant. For example, during a singular engagement, combat exists to prevent opponent interference with a mission through the use of deadly force [1]. The use of deadly force was borne of the utility that comes from permanent threat removal coupled with the lack of an effective means of reversible incapacitation. However, what if the operation could be completed without a need for deadly force? The objective would still be achieved but without the permanent harm inherent in conventional weapons. It is necessary to note that no party investigating the use of NLW suggests that they could ever completely replace lethal systems. Rather, NLW will complement conventional munitions and introduce a degree of graduated lethality and capability previously missing from engagements where only lethal means were available.

Therein lies one of the ethical issues with the development and deployment of directed-energy counter-personnel NLW: that such systems are actually feasible battlefield weapons or can be in the near future. Existing NLW either don't affect individuals directly (counter-materiel) or can only reasonably affect individuals (Taser) or small groups (tear gas) for each deployment. However, new architectures can affect larger groups, over a larger range, and with faster deployment times. The prototype Active Denial System architecture uses millimeter-wave radiation to cause a burning sensation in its multiple targets [36]. Regardless of the weapon specifics, these technological improvements are arguably on par with the evolution from single-shot muskets to the machine gun. Both are weapons, which fire projectiles using explosives, but one hits a singular target at a low firing rate and the other is a localized weapon of mass-carnage. Similarly, the ADS is effectively an amplified Taser in that pain and incapacitation can be inflicted at longer ranges and to multiple targets simultaneously.

The change in NLW architecture to a true *weapon* rather than a *control tool* is what has garnered such an ethical backlash from organizations such as the American Red Cross and Human Rights Watch [2]. Other groups such as the ACLU argue against the application of NLW and claim that "the road to hell is paved with [...]" good intentioned Non-Lethal Weapons [3]. In a more specific scenario, the UN Committee Against Torture identified Taser and Taser-like NLW as almost a form a torture and advocated against their continued use [4]. Overall, some scientists, such those at Federation of American Scientists, suggest that NLW not only cause unethical levels of pain but also that the systems are not useful in the first place [3]. As mentioned previously, new counter-personnel NLW are effectively weapons that *simulate* rather than *inflict* harm. The major ethical question then arises, "is it better to harm or to kill?" The secondary question is "what are the effects of collateral damage when evaluated against this harm?"

It can be reasonably inferred that in the conduct of human interpersonal interactions, violence is abhorred, especially murder. While ethicists have sought to contextualize and contrast the gravest sin of *murder* with the state-sanctioned killing of military combat, it remains true that minimizing casualties is to be encouraged by the parties involved [35]. This is doubly so for civilian *collateral damage*. The prevention of civilian harm is among the most important bedrocks of both the Laws of War (Geneva Conventions, arms control treaties, etc.) and the "codes of war" enshrined by professional militaries [35].

The most prevalent argument against Non-Lethal Weapons hinges on the ethics of employing such systems against both combatants and any innocents that might be caught in the area of effect. Primarily, these points rely on the assumption that the infliction of pain is a greater ethical evil than conventional killing, maiming, or injury of these parties. These issues also extend to protect the soldiers operating the equipment. This is especially true since directed-energy area-affect NLW have an almost 100% rate of effect on targets in the area of interest (though one target might not be as vulnerable as others), while conventional munitions do not. As a result, the operator of an NLW knowingly inflicts pain or other short-term effects on civilians in the area of effect, while the operator of conventional munitions could almost plead ignorance in this regard.

Supplementing the explicit laws of war are more nebulous "codes of war" which harken back to chivalry and similar unofficial guides for military conduct. These codes vary from culture to culture and have two primary goals. The first is to ensure that the harm to vulnerable peoples (prisoners of war, civilians, aid workers, etc.) is minimized. The second is to protect the combatants from engaging in activities that are morally or ethically haunting. Shannon French, in her book *The Code of the Warrior*, focuses on the latter element of "codes of war." She describes the need for combatants to consider themselves as though they are "fighting the good fight" and that they are not morally or ethically sullied by what happens in wartime [35]. Moral, ethical, and emotional tolls can affect combatants dramatically. Syracuse University even has a dedicated "Moral Injury Project" in which researchers seek to understand and raise awareness about individuals coping with actions that inflicted a mental toll on their well-being [37]. Moral injuries are traditionally grouped with post-traumatic stress disorder, but they are usually more explicitly linked to decisions and actions taken by the individual affected rather than a long-term compounding effect. Morally injurious actions include wounding or killing civilians accidentally, giving orders in combat that result in the death of allies or subordinates, and any news about actions taken against allies and collaborators in the area of operation. NLW promise to address one of the most common sources of moral injury: that of inflicting permanent or deadly harm on civilians.

It is at this juncture that directed-energy Non-Lethal Weapons are introduced. Some of the newer NLW have an area-effect similar to larger explosives, but notably minimize the long-term effects on those impacted. In the case of one-for-one replacement of conventional area-effect munitions, NLW are the more ethical choice due to minimal permanent harm. However, NLW promise to achieve new architectures of engagement precisely because their effects are temporary and reversible. For example, NLW could be used in hostage situations, incapacitating all people in an area, both innocent and combatant. With the combatants safely incapable of violent response, forces could move into apprehend the targets and rescue the hostages. However, this mission success comes at the cost of the NLW operator knowingly affecting, albeit temporarily, the innocent civilian hostages. This complicates the ethical arguments regarding NLW by effectively condoning violence against civilians, an action that has been discouraged and outlawed by both the legacy "codes of war" and international agreements such as the Geneva Conventions.

When it comes to the use of weapons or conduct in wartime, the "Doctrine of Double Effect" (DDE), as best discussed by theologian Thomas Aquinas, is of significant interest [38]. More specifically, the DDE is applicable to cases of collateral damage. This ethical principle is essentially a more limited form of "the ends justify the means." However, for the DDE, the negative effect cannot ever be caused intentionally. For example, killing with intent to kill an assailant is incompatible. However, killing an assailant when intending purely self-defense is more acceptable. As outline by *The New Catholic Encyclopedia*, the DDE has four tenets or elements (quoted below) [39]:

(1) "The act itself must be morally good or at least indifferent."
(2) "The agent may not positively will the bad effect but may permit it. If he could attain the good effect without the bad effect he should do so. The bad effect is sometimes said to be indirectly voluntary."
(3) "The good effect must flow from the action at least as immediately (in the order of causality, though not necessarily in the order of time) as the bad effect. In other words the good effect must be produced directly by the action, not by the bad effect. Otherwise the agent would be using a bad means to a good end, which is never allowed."
(4) "The good effect must be sufficiently desirable to compensate for the allowing of the bad effect."

Looking through these four elements in turn, the DDE can be applied to different weapons systems or mission architectures. As a case study, consider the scenario of a precision bombing attack against a terrorist leader:

(1) The incapacitation of the terrorist leader, who ordered attacks against civilians, etc. (proven record of immoral acts) is a moral good and meets the first criterion.
(2) The precision bombing may also hurt innocents near the terrorist leader, but they are not the primary target—this meets the second criterion because while the death of innocents is a "bad effect" and anticipated, it is unintended.
(3) The good effect arises from incapacitating the terrorist leader, not from harming the civilians nearby—this meets the third criterion.
(4) The good effect of incapacitating the terrorist leader outweighs the collateral damage due to the future harm said leader can and would inflict—this meets the fourth criterion.

As a counterpoint case study, the DDE can be similarly applied to a terrorist bombing against a civilian populace:

(1) In the mind of the bomber, the overall act of weakening the enemy's resolve is a moral good—this meets the first criterion.
(2) The bomber aims specifically to kill civilians (the "bad effect") violates the second criterion.
(3) The act of killing civilians is the way to weaken the enemy's resolve – this violates the third criterion.
(4) In the mind of the bomber, the weakening of the enemy is worth the price of those innocent lives—this meets the fourth criterion.

As straightforward as the DDE is overall, evaluating these four criteria in turn is controversial. For the US, the first and third criteria are easily met due to extensive doctrine, training, and other cultural efforts. For the sake of this analysis, US intentions are assumed to be good (meeting the first criterion). Soldiers are also trained and never asked to fire on or otherwise directly harm civilians (meeting the third criterion). The most debatable of the four criterion though, are the second and fourth elements. The second criterion regards the intentions and specific course of action taken by the soldier, combatant, etc. The fourth criterion regards the "cost-benefit analysis" of the overall activity.

Non-Lethal Weapons run afoul of the second criterion because, as mentioned previously, these systems (usually) have confirmed effects on all people within their engagement area. Therefore, the action of using a NLW, such as the Active Denial System, apparently violates the second criterion of the DDE due to the knowledgeable infliction of harm on innocents. However, this is balanced by the unwilling application of harm on those affected civilians. Furthermore, applying such principles to NLW in isolation misses the necessary comparison with current conventional means to accomplishing a mission. The choice is not between using a NLW and leaving civilians untouched, but it is between using a NLW and using conventional munitions or other engagement techniques. A NLW will have fewer long-term repercussions as compared to a traditional firefight due to minimization of permanent casualties. While a fortunate conventional engagement may leave no civilians directly impacted, such a scenario is becoming increasingly unlikely given the tactics and techniques of various non-state actors such as the Islamic State [3]. When the argument is thus rephrased between current conventional and near-future NLW options, NLW are readily the more ethical solution. However, this necessitates the assumption that the engagement will occur regardless.

The fourth criterion of the DDE addresses the so-called "cost-benefit analysis" implicit in any combat operation. More specifically, "the good effect must be sufficiently desirable to compensate for the allowing of the bad effect." However, how can one measure the magnitude of the relative effects? The most readily available comparison is between death and continued, unimpeded life. Regarding the latter, this chapter specifically avoids a detailed discussion regarding the debate between death and a continued, low-quality of life survival by an individual. Indeed, that comparison is what is unlocked by the employment of NLW. Assuming that such systems, after appropriate testing and development, have no long-term consequences, the number of civilian casualties should decrease dramatically in situations when conventional munitions would otherwise be employed. Currently, traditional engagement techniques may lead to injured civilians whose quality of life are impeded by their injuries, while NLW specifically avoids the situation entirely while minimizing the number of civilians permanently impacted. As a result, NLW presents the best opportunity to meet the fourth element of the DDE by further reducing the magnitude of any "bad effects" arising from a combat engagement.

In short, the Doctrine of Double Effect reinforces the view that NLW are the superior option for military operations. NLW minimize the "bad effects" of combat and are no more unethical than conventional munitions when assessing the second criterion of the DDE. Furthermore, due to the inherent reversibility and impermanence of NLW, their application in lieu of conventional (and risking lethal) force is the ethically superior option. This counters the logic that NLW effects are somehow less ethical or permissible than lethal force and this is especially true when taken in the context of preventing harm to other civilians from unchecked opponent activity. With regards to increasing military capability, NLW reduce the repercussions of combat operations, and therefore they decrease the threshold for what constitutes a "good effect." With a lower "good effect" requirement, there are more targets available to military planners without compromising the balancing act necessitated by DDE, assuming NLW can be employed. Thus NLW are both more ethical than conventional munitions and enable more opportunities for ethical actions to ensure national security.

9 Closing Thoughts

> Traditional weapons offer two difficult choices – do nothing or kill. Non-Lethal Weapons (NLW) offer alternatives.—John Alexander (Colonel, United States Army, Retired) [3]

As humanity moves forward in an increasingly globalized society, technological development and proliferation is a fact of life. It is fallacious to assume the costs of such developments universally outweigh their benefits. However, the international community has learned from past conflicts and, as such, has agreements, treaties, and bodies that attempt regulate the development, use, and proliferation of technology. It is also of note, though, that such agreements usually do not exist until the technology affected presents a clear risk to innocent people or has the potential to disrupt the international order. Until these agreements are negotiated, further development of directed-energy Non-Lethal Weapons should proceed with the utmost urgency.

As revolutionary as NLW are, one cannot deny that such systems will have a mix of positive and negative effects at large. However, focusing on the costs or risks before a full understanding of the positive effects can constrain technological and ethical advancement. Fear-mongering risks not only impeding technological progress, but it also impacts the United States' ability to counter such technologies.

As a result of this change in architecture, current international agreements do not specifically limit the use or development of novel NLW.

References

1. Giri DV (2004) High-power electromagnetic radiators: nonlethal weapons and other applications. Harvard, Cambridge, MA
2. Anderson G (2016) How the red cross killed non-lethal weapons. Wall Str J, 22 June 2016
3. Alexander J (2007) An overview of the future of non-lethal weapons. Med Confl Surviv 17(3):180–193
4. Nizza M (2007) U.N. torture panel singles out tasers. The New York Times, 26 Nov 2007
5. DoD Executive Agent for Non-Lethal Weapons (NLW), and NLW Policy, 3000.03E. United States Department of Defense (2017)
6. Lewer N (ed) (2002) The future of non-lethal weapons: technologies, operations, ethics, and law. Frank Cass, London
7. Hambling D (2005) Maximum pain is the aim of new US weapon. New Scientist, 2 Mar 2005
8. Kusher D (2013) The real story of stuxnet. IEEE Spectrum, 26 Feb 2013. https://spectrum.ieee.org/telecom/security/the-real-story-of-stuxnet/
9. Maslow AH (1966) The psychology of science. Harper & Row, New York
10. A marine general at war. Slate, updated 22 April 2010. http://www.slate.com/articles/life/risk/2010/04/a_marine_general_at_war.html
11. XM395 Precision Mortar (2018). Accessed 20 April 2019. https://www.northropgrumman.com/Capabilities/GuidedProjectiles/Documents/xm395PGKForMortarsFactSheet.pdf
12. The Editors of Encyclopaedia Britannica (2016). Carpet bombing. In: Encyclopaedia Britannica. https://www.britannica.com/topic/carpet-bombing
13. Watts B (2013) The evolution of precision strike. Center for Strategic and Budgetary Assessment. https://csbaonline.org/research/publications/the-evolution-of-precision-strike
14. Krechetnikov A (2012) Moscow theatre siege: questions remain unanswered. BBC News, updated 24 Oct 2012. https://www.bbc.com/news/world-europe-20067384
15. Solid state laser testbed. United States Army Space and Missile Defense Command. https://www.smdc.army.mil/FactSheets/SSLT.pdf
16. Tzu S (c. 400 BCE) The art of war (trans. Samuel B. Griffith). Clarendon, Oxford
17. Tillerson R (2017) Remarks to the NATO-Ukraine Commission. https://nato.usmission.gov/march-31-secretary-state-tillersons-remarks-nato-ukraine-commission/
18. MH17 Ukraine plane crash: What we know. BBC News, updated 28 Sept 2016. Accessed 13 April 2019. http://www.bbc.com/news/world-europe-28357880
19. Ingersoll G, Szoldra P (2013) 19 Unforgettable quotes from retiring general James 'Mad Dog' Mattis. 23 Jan 2013. https://www.businessinsider.com/general-maddog-mattiss-best-quotes-2013-1
20. Loeb V (2002) Bracing for 'Primoridial Combat. The Washington Post, 31 Oct 2002
21. History. Non-Lethal Weapons Program, United States Department of Defense. Accessed 13 April 2019. https://jnlwp.defense.gov/About/History/
22. Geneva Conventions and Commentaries (2014). Accessed 13 April 2019. https://www.icrc.org/en/war-and-law/treaties-customary-law/geneva-conventions
23. Prange G (2001) At dawn we slept: the untold story of Pearl Harbor. Penguin, New York
24. Schwalbe S (2016) Where are the terrorist WMD attacks? Updated 7 Oct 2016. Accessed 13 April 2019. http://inhomelandsecurity.com/where-are-the-terrorist-wmd-attacks
25. Chemical weapons convention. Accessed 13 April 2019. https://www.opcw.org/chemical-weapons-convention/
26. Savage C (2009) Videos shed new light on Sotomayor's positions. The New York Times, 10 June 2009
27. Fidler D (2007) 'Non-lethal' weapons and international law: three perspectives on the future. Med Confl Surviv 17(3):194–206
28. Geneva Conventions and Commentaries. Article 3. https://ihl-databases.icrc.org
29. Chemical Weapons Convention. Article 2. https://www.opcw.org/chemical-weapons-convention/articles/article-ii-definitions-and-criteria

30. Protocol on Blinding Laser Weapons. Accessed 13 April 2019. https://ihl-databases.icrc.org/ihl/INTRO/570
31. Distributed sound and light array. Non-Lethal Weapons Program, United States Department of Defense. Accessed 13 April 2019. https://jnlwp.defense.gov/Press-Room/Fact-Sheets/Article-View-Fact-sheets/Article/577990/distributed-sound-and-light-array/
32. Scott R (2010) War without death? How non-lethal weapons could change warfare. Christian Science Monitor, updated 11 March 2010. Accessed 13 April 2019. https://www.csmonitor.com/Commentary/Opinion/2010/0311/War-without-death-How-non-lethal-weapons-could-change-warfare
33. Slippery slope. Texas State Department of Philosophy. Accessed 13 April 2019. http://www.txstate.edu/philosophy/resources/fallacy-definitions/Slippery-Slope.html
34. Callway E (2017) Second Chinese team reports gene editing in human embryos. Nature, 8 April 2017. https://www.nature.com/news/second-chinese-team-reports-gene-editing-in-human-embryos-1.19718
35. French S (2004) The code of the warrior: exploring warrior values past and present. Rowman & Littlefield
36. Active denial technology factsheet. Non-Lethal Weapons Program, United States Department of Defense. Accessed 13 April 2019. https://jnlwp.defense.gov/Press-Room/Fact-Sheets/Article-View-Fact-sheets/Article/577989/active-denial-technology/
37. Moral Injury Project. http://moralinjuryproject.syr.edu
38. McDonald W (ed) (2002) The new catholic encyclopedia. McGraw-Hill
39. Doctrine of Double Effect. Updated 24 Dec 2018. http://plato.stanford.edu/entries/double-effect

Disruptive Energy Technologies and Military Capabilities

Anmol Soni

Abstract Energy is a critical input in military functions. As more advanced technology and weapons are deployed, the demand for energy is also expected to rise. However, it is pertinent to examine the possibility and extent of any fundamental changes in the way energy technology is deployed by the US military and the implications on national security. While large research projects are currently underway to support moves toward enhancing energy efficiency in operational energy usage, it is unclear to what extent these would create major tactical or strategic disruptions. To identify the possibility of major energy technology related disruptions, an exploratory framework that analyzes different options along the different stages of the energy value chain—generation, storage, transfer, and usage—is developed, and different technologies that fit within it are identified. These technologies are then assessed in the context of the Lethality-Survivability-Mobility triad framework to understand where and how strategic disruptions could be generated. Finally, this chapter provides recommendations on key areas where research could be pursued to gain maximum returns and the organizational aspects of energy technology acquisitions of the US military.

1 Introduction

Energy plays a critical role in supporting the defense establishment in both combat and non-combat settings [1]. In combat, the ability to power the forces continuously and flexibly can play an important role in determining the outcome of conflict. For example, the ability of a soldier to move unencumbered by batteries and the need to fuel forward operating bases (FOBs) can provide comparative advantage by enabling longer duration missions. In non-combat situations, as well, the resilience of energy systems is crucial to ensure efficiency, security, communications, and medical response. The military's energy strategy is undergoing a change in response to the rising pressure on resources and the changing capabilities and types of technology

A. Soni (✉)
Georgia Institute of Technology & Georgia State University, Atlanta, GA, USA
e-mail: anmolsoni@gatech.edu

M. E. Kosal (ed.), *Disruptive and Game Changing Technologies in Modern Warfare*, Advanced Sciences and Technologies for Security Applications,
https://doi.org/10.1007/978-3-030-28342-1_7

available. Further, the high dependence on petroleum exposes military's energy costs to volatility in global oil prices. Responding to pressures to cut spending, there is a need to enhance efficiency and seek alternative sources of powering the military [2]. Lastly, with national level military and other strategic documents being rebalanced toward near-peer competitors [3] and the Asia-Pacific [4], the demand for energy is likely to witness a shift in its structure and composition. The strategic rebalancing affects both the type of fuels needed and the technologies that consume energy. This provides a useful setting to examine the role of energy in national security, the extent to which its nature and composition is likely to change in the near to long term future, and whether and to what extent can we expect a disruption in energy technologies used by the military.

Energy usage in the military is categorized into Installation Energy and Operational Energy, where the former includes consumption of energy at the domestic bases, and the latter is defined as "the energy and associated systems information and processes required to train, move and sustain forces and systems for military operations" (10 US Code § 2924) (US Code, Title 10). The US Department of Defense (DOD) has set a goal of increasing the share of renewable energy in its total energy consumption. In fact, legislation mandates that the share of renewable energy in the total facility energy consumption, which includes *building, installation, structure, and other property owned by the Federal Government*, to be at least 25% by 2025. However, the goals do not specify any targets for operational energy consumption, i.e., the energy used by forward operating bases (FOB), ground and air vehicles, tactical centers, and soldiers in combat.

The Operational Energy Strategy [5] released by Office of the Assistant Secretary of Defense for Energy, Installations, and Environment outlined three goals:

- Increase future warfighter capability
- Identify and reduce risk
- Enhance current mission effectiveness

Each of these objectives affects different aspects of the energy value chain and has varying implications for all types of energy use, whether ground troops, vehicles, aircraft, unmanned systems, or forward operating bases (FOBs). This chapter examines the drivers and factors affecting consumption of energy operationally by the US military and explores the possible disruptive energy technologies likely to emerge based on the current research. It is important to note that the analysis is for the operational energy usage of the military and does not directly address the military bases and installation energy consumption within the US; it is focused on overseas operations of strategic significance.

2 Motivation and Relevance

Between the growing emphasis on reduced spending and the constantly widening base of threats, energy usage by the military is assuming a critical position. The high dependence of the forces on energy and the role of energy in determining the tactical decisions taken by the US military add to the complexity of this issue [6], tactical decisions that may have broader effect on operations. Research on energy technologies has been identified as a one of the priority areas in the Department of Defense Science and Technology Communities of Interest [7]. As the forces adopt more advanced equipment and technology, the demand for energy is increasing [8]. Research agencies, inside and outside the DOD, are making advances to harness alternative sources of energy and enhance the energy efficiency of existing weapons, vehicles, and equipment. This creates a need to put in place fundamentally dynamic systems that can respond to the changing nature of conflict and rapidly deploy newly available sources of energy and related technology. Energy usage by the military has operational and strategic implications on national security. For policy, the types of fuels used, forces deployed, and long-term posturing are all affected. Location of bases, length of operations, and the ability to cover large areas during patrols or expeditionary deployment affect the military's operational strategy. Energy is a major component of these logistic factors and related decision making processes.

A disruption in the energy technology used by military has the potential to cause a fundamental change in the way energy is produced, transported, and consumed, necessitating a major overhaul of the system of energy flows in the individual, vehicle, squad, and base usage. It is therefore important to examine the energy-military linkages, the advancements being made, and their ability to fundamentally disrupt present-day and future operations.

This chapter conducts a systematic analysis of available and likely future technologies along the energy value chain, and it then explores their ability to cause disruption based on the Lethality-Survivability-Mobility triad. Most available analyses are focused on technologies and their implications. However, a value-chain based framework approach is adopted in this chapter to identify the critical areas where disruptions can take place. The focus is on examining the systemic approach to developing energy technology in military operations rather than the usual approach of looking at energy as a component of technical logistics. This is cognizant of the view that the process of decision making is rooted deep in program design, funding of research, and fundamental organizational approach to technology and not just the scientific role thereof. The analysis in this regard is based on the acquisition and logistics programs of the military and the areas where the process can be leveraged or improved to increase the pace at which energy related technology can be adopted in forward operations.

3 Background

Operational energy continues to be high, even though it has been stable or declining in the recent past (Fig. 1). In FY2011, the US DOD used the equivalent of 115 million barrels of oil in operational consumption and in FY2016, the consumption was 85 million barrels at a cost of nearly $9 billion (USD). The dominant consumer was the Air Force, followed by the Navy. Within the Air Force, air-lift and fighters comprise

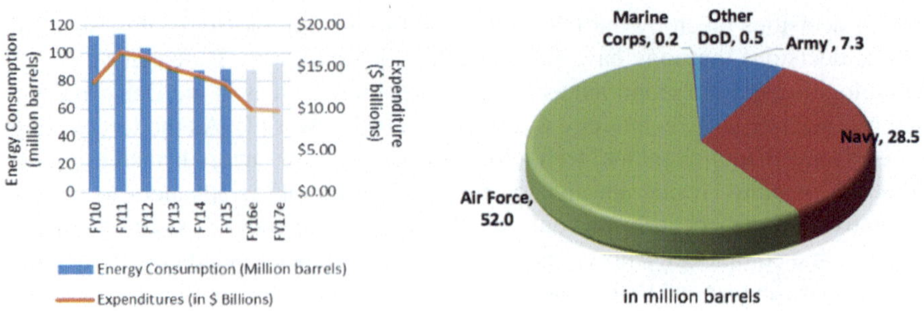

Fig. 1 DOD operational energy consumption: trend and composition (FY 2015). *Source* Data from Operational Energy Annual Report FY 2015

the bulk of operational energy consumption. Conventional, petroleum-based, fuels form a bulk of this energy.

The logistics of transporting the fuel, purchasing it on the international markets close to the source, and the vulnerability caused due to the dependence on these sources of energy play a particularly important role in determining the way FOBs operate, the ability of ground troops as measured by mobility, the energy used by combat troops, and the overall performances of the forces in situations of conflict. For example, reducing the load of batteries carried by combat soldiers during conflict affects the distance they can cover, the duration for which they can operate without refueling, and the type of operations they can execute [6]. In the case of FOBs, a similar reduction in fuel usage due to improvements in efficiency and possibly an altogether removal of dependence on fuel supplies from outside because of increased generation using renewables at the base can be instrumental in changing operational energy consumption patterns. This has implications not just in terms of cost savings but also lives saved in the avoided fuel transportation missions [1, 9].

4 Framework and Methodology

A detailed analysis of publicly available sources was conducted to identify and describe the key energy technologies being developed or adopted in different operational applications in the US military. These new and emerging technolgoies were then analyzed along the energy value chain, covering upstream, midstream and downstream energy use, including energy generation/production, storage and transfer, and usage. However, for a technology to be truly disruptive, this chapter investigates variables that can cause a "paradigm shift," that is a change in the way military operations are conducted, like the shift from horses to tanks The iron-triad of Lethality-Survivability-Mobility framework, depicted conceptually in Fig. 2, is useful in facilitating a first assessment of the available technology options along these lines [10]. The triad interacts to determine key choices and decisions that the mili-

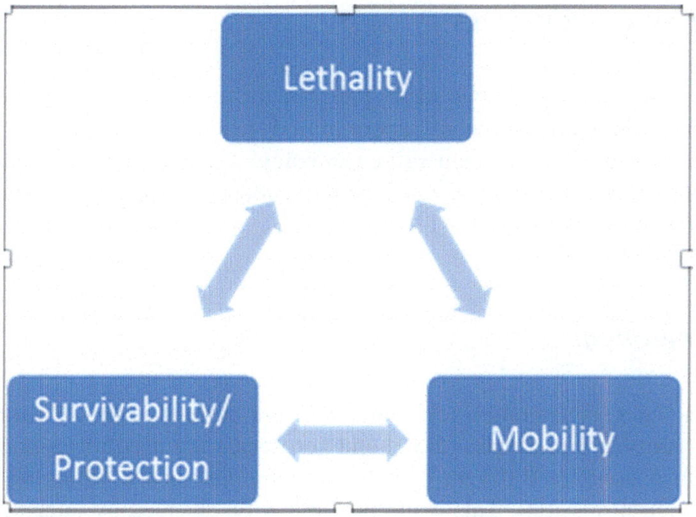

Fig. 2 Framework for analysis

tary makes. Acquisition and deployment of technology is a balance among the three elements of the triad. Historically, survivability is often achieved via added mass as protection in the form of shielding or armor, sometimes to the detriment of mobility.

Before proceeding, it is pertinent to provide a working definition of these three dimensions. Lethality of a technology would provide an indication of its ability to cause damage and impair the functioning of opponents. A more formal definition is found in the ADP 3-0, Unified Land Operations document:

> The capacity for physical destruction is fundamental to all other military capabilities and the most basic building block for military operations. Army leaders organize, equip, train, and employ their formations for unmatched lethality under a wide range of conditions. Lethality is a persistent requirement for Army organizations, even in conditions where only the implicit threat of violence suffices to accomplish the mission through nonlethal engagements and activities. The capability for the lawful and expert application of lethal force builds the foundation for effective offensive, defensive, and stability operations. (ADP 3-0) [11]

Survivability or protection would be the ability to withstand shocks and continue to perform in adverse conditions as well and is defined as "All aspects of protecting personnel, weapons, and supplies while simultaneously deceiving the enemy." (JP 3-34) [12]. The third component, mobility, refers to the possibility of moving from one location to another without being encumbered: "A quality or capability of military forces which permits them to move from place to place while retaining the ability to fulfill their primary mission." (JP 3-17) [13].

Based on the findings from the value chain and Lethality-Mobility-Survivability analysis in this framework, the specific developments and/or applications that need to be monitored for a possibility of causing major disruptions in combat applications are explored.

5 Analysis and Findings

The findings are presented in two stages based on the methodology described above. The possible technology options are superimposed on the energy value chain. Figure 3 summarizes this and notes key emerging technologies that have been identified. The energy value chain starts at the energy generation node, covering conversion, storage, transfer, and ending at usage.

5.1 *Generation*

Most of the energy generation for operational use currently takes place at FOBs. Typically, petroleum products—diesel and kerosene—are used to generate power, which is then used for different operational activities such as charging batteries to support patrol missions. This could potentially be changed by integrating more renewable sources in the energy mix—allowing patrol missions to generate power even when on the move rather than depending primarily on batteries.

5.1.1 Energy Generation at Fixed Locations

Most of the energy used at overseas bases is purchased from the international markets [14]. While efforts are made to minimize the distance between the points of purchase and usage, this still requires significance transportation capacity, is expensive, and creates logistics burdens. A fundamental change in the process of energy generation can be enabled if FOBs are able to generate their own energy without

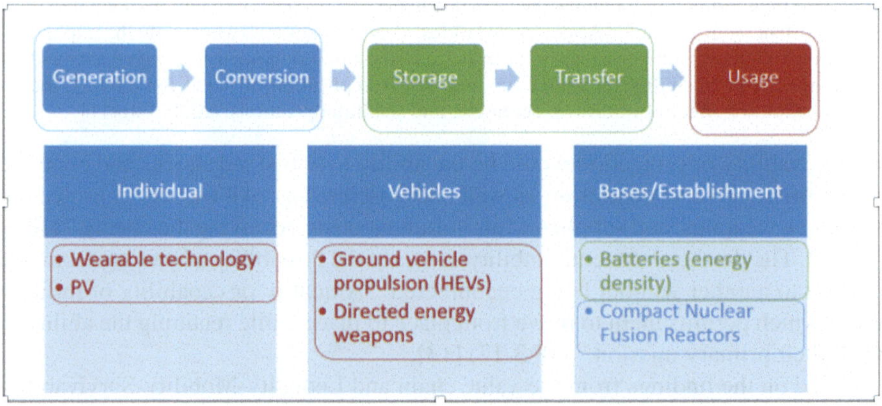

Fig. 3 Mapping key emerging technologies on the energy value chain. Note the color coding of the technology refers to the stage in the energy value chain affected by the technology

depending on conventional fossil fuels. This is occurring through the increased usage of renewable sources of power [6, 15]. For example, the DOD's Operational Energy Strategy includes the goal of increasing usage of alternative fuels and harvesting energy at the point of use to minimize the burden and associated risk of resupplying liquid fuels. Further, in 2017, experts from the Army, Navy, Air Force, and Defense Logistics Agency (DLA) are conducting testing and evaluation for alternative fuels that can easily be mixed with commercial fuels in hard to reach operational areas [16].

However, a critical requirement for power generation is limiting vulnerability due to visibility, i.e., these power sources should not become a visible target for the opponents/enemies. This should not create an added vulnerability since oil barrels and generators are existing vulnerable targets. Additionally, the ability to install new energy sources in limited time-frame and area is important. For example, it would be difficult to harness wind energy easily due to both the cost of setting it up as well as the visibility of turbines. However, photovoltaic (PV), solar-based, technology applications in the bases in the Middle East and western Asia have been successful. An example of this is the Ground Renewable Expeditionary Energy Network System (GREENS) [17]. The modular solar panels allow flexibility and ease of setting-up in bases. These are also smaller (and therefore more mobile) compared to wind based power generation and therefore have low visibility.

Another example of in-development source of energy generation is fusion-based compact nuclear reactors. This technology is still in the early phases of development and will perhaps not be available for deployment in the near to medium term. Lockheed Martin is reported to be developing such a capability, which would be safe and limited, as the reaction would be contained in a magnetic chamber [18]. Since then, the company had registered a patent for the compact fusion reactor [19]; however, any information about specific dates of the project and implementation of pilots is not available. The University of Washington is also reported to be working on an ARPA-E grant to develop nuclear fusion reaction capability [20]. In 2015, CT Fusion, a spin-off company from a University of Washington project, continued to work on this area. The key challenge that the teams face is scaling up and stabilizing the reaction. However, there is no clarity in the publically available accounts on the size and space of this project. Therefore, while this technology presents a possibility of disrupting the way energy is generated and secured in forward military operations, its timeline is unclear. Further, the political challenges and negative perceptions associated with nuclear power and the response from local communities and ground forces might be a cause of concern in implementing any nuclear related technologies at the base level.

5.1.2 Energy Generation by Individuals

Reducing the load carried by the individual soldier or marine is major challenge for the 21st Century. It's estimated that average kit for a soldier deployed could be as much as 130 lb which is much higher than the prescribed 33% of body weight [21].

Another estimate puts this weight at closer to 200 lb [22]. These heavy loads not only reduce warfighting abilities but can also lead to chronic health problems for soldiers in the long term [23]. Batteries comprise a large share of this burden and can add anywhere between 20 and 25 lb depending on the length of the mission and type of equipment carried; these numbers are expected to increase in the coming years due to the energy requirements of the gear that soldiers carry [24].

If ground troops could seamlessly produce energy via walking or other activities, that weight burden could be lessened. In addition to the application of PV for power generation at fixed locations, another particularly useful application of PV technology is through the integration of photovoltaic fibers to power individual soldier. Called electro-textiles, these smart materials contain fibers that generate electricity, which can either be used to power devices or charge batteries, thereby significantly reducing the load of warfighter [9, 21]. This would allow troops to produce power as they move in the battlefield, reducing the battery load they are required to carry while also increasing the potential time away from a base. Another example of such technology is the development of piezoelectric materials that directly convert mechanical energy, such as from the impact of a boot heel during walking or other ground-based movement, to electro-chemical energy. Researchers at Georgia Tech have led investigation into the creation and development of such materials [25].

5.1.3 Storage and Transfer

Storage and transfer of energy form the next stage in the energy value chain. This section explores the role and evolution of batteries, especially those used by ground troops. Batteries comprise a substantial portion of the load carried by soldiers [2]. However, among batteries, there exists a significant variation ranging from compact batteries that soldiers carry to large storage systems used to power and stabilize energy availability for vehicles or at a fixed base. For portable batteries, the aim is to reduce the weight carried by soldiers and the improve survivability during the entire length of the mission. The primary technology currently in use at the individual ground troop level is based on Lithium-ion (Li-ion) batteries. For Afghanistan missions, batteries are a significant part of the total weight carried by ground troops. Soldiers often carry seven types of batteries that can weigh a total of 20 lb [26].

Harnessing fuel cell technology is another route that may offer solutions [27]. In such technology, chemical fuel is directly converted to electric power. As renewable energy is integrated into power generation, managing surge generation and balancing supply and demand will be needed. For easing the load of ground troops, advances are being made in ergonomically designed conformable battery storage. Ensuring that the battery fits with the rucksack and provides a fightable form factor is necessary. Additional research is underway exploring ways to increase the energy density of batteries, thereby to increasing the energy obtained per unit of weight carried. This reduction of load and improvement in flexibility enhances the agility of the troops.

One example of wearable fuel cell-based prototype cartridges has been developed using aluminum hydrides (AlH_3) as a fuel source [28]. These may provide significant advantage in terms of the weight compared to the existing solution for Li-ion batteries. In addition, two other types of fuel cells currently being examined are alkaline-based fuel cells, for stack or system usage and solid oxide fuel cells, for replacing generators [9].

Systems-based solutions holistically address the energy production-consumption value-chain. An example is the Soldier Wearable Power System (SWIPES) that was selected as one of the top 10 US Army Greatest Inventions in 2010. The system powers soldiers for extended mission times by integrating force protection, electronics, and communication equipment with advanced power sources [29]. The conformable battery system allows for "individual tailoring for the soldiers," and reduces the battery weight by 30–50% [29].

5.2 *Usage*

5.2.1 Vehicles

Considering the high share of energy usage in vehicles in military operations, increasing the efficiency of vehicles and eventually even moving away from the consumption of fossil fuels will help reduce this burden [6]. Hybrid vehicles, improved ground propulsion, and blending of renewable fuels offer potential alternatives. For instance, the US Navy, in pursuance of its Great Green Fleet initiative, is increasing the share of alternative fuels in its total energy consumption portfolio [30]. However, renewable fuel purchases would have to meet the requirement of the current fleet and its engine and technology design requirements. For the US Army, the Army Tank Automotive Research, Development and Engineering Center (TARDEC), in collaboration with the Vehicle Technology Office of the Department of Energy, has created the Advanced Vehicle Power Technology Alliance to work on the improving fuel economy for the ground tactical fleet program. The key focus areas of this program are automation/smart cruise control, auxiliary electrification, thermally efficient cylinders, and modeling and simulation for reduced vehicle weight [31]. In 2017, tests to demonstrate the feasibility of hydrogen fuel cell technology in vehicles have been conducted [32]. Done in collaboration with General Motors, the program aims to leverage hydrogen power for vehicles, which will reduce the dependence on petroleum for operations since the vehicle fleet is predominantly dependent on petroleum-based fuels and hydrogen would be a substitute.

5.2.2 Weapons

Finally, an emerging application of energy in military is the use of directed energy in non-lethal weapons. Another energy-intensive area of research is for non-lethal weapons. Nonlethal weapons provide the military with options to escalate force but minimize casualties and collateral damage, and will effectively fill the gap on the escalation of force continuum [33]. These weapons help reduce own force and civilian casualties [34]. In particular, directed energy applications include Radio Frequency and High powered Microwave Systems to deter vehicles, and sound and light based technology to suppress and disrupt individuals [35]. Another example is the Active Denial System (ADS), a microwave ray that causes intense superficial heating. It is intended for use to protect structures, perimeter security, and riot control situations [36].

Directed energy weapons, often referred to as rail guns, mounted on vehicles are an example of applications being explored for future use in against an adversary's armored vehicles and for anti-access and area denial (A2AD). For example, the Ground Based Air Defense (GBAD), being developed by the Office of Naval Research, mounts a high-energy laser on a ground vehicle [37]. An on-board system locates unmanned aerial targets and passes the information to the command, control, and communications platform, which locates and tracks the target [38].

In summary, dozens, if not hundreds, of programs and projects have been undertaken by different parts of the US defense establishment, its associated research organizations, and the private sector to support the development of energy technologies for usage in military operations. The broad underlying objectives of these are enhancing fuel efficiency, supporting the warfighter in the battlefield, and incorporating alternative means of energy production at the base, squad, and individual levels.

6 Lethality-Survivability-Mobility Assessment

Based on this initial identification of the key technologies that have a potential to affect the way energy is produced and consumed in combat, Table 1 provides an assessment of these technologies using the Lethality-Survivability-Mobility framework introduced earlier. The assessment is conducted for three time periods: immediate applications, development in the medium term (approximately 5 years), and in the long term (10 years or more). The technologies are ranked on a scale of Low, Medium, or High. Technology applications that are not likely to be adopted in the

Table 1 Analyzing the technology options on the lethality-survivability-mobility framework

			Current			Medium Term (in 5 years)			Long term (in 10-15 years)		
			Lethality	Survivability/ Protection	Mobility	Lethality	Survivability/ Protection	Mobility	Lethality	Survivability/ Protection	Mobility
Operational Energy	Individual	Wearable Technology	Poor	Marginal	Marginal	Marginal	High	High	Marginal	High	High
		Photovoltaics	Poor	Marginal-Poor	Marginal	Poor	High	High-Marginal	Poor	High	High
	Vehicles	Ground Vehicle	Marginal	High-Marginal	Marginal	Marginal	High-Marginal	High-Marginal	High-Marginal	High-Marginal	High-Marginal
		Directed Energy	Marginal-Poor	High-Marginal	N/A	Marginal-Poor	High-Marginal	N/A	Marginal	High-Marginal	N/A
	Forward Operation Bases	Batteries	Poor	Marginal	High-Marginal	Marginal-Poor	High-Marginal	High-Marginal	Marginal-Poor	High	High
		Compact Nuclear Fusion Reactors	N/A	N/A	N/A	N/A	N/A	N/A	High-Marginal	Marginal-Poor	High
Institutional Energy	Fixed Bases		Poor	Poor	Poor	Poor	Poor	Poor	Poor	Poor	Marginal

Legend

High	High-Marginal	Marginal	Marginal-Poor	Poor	N/A

time periods or provided minimal marginal improvements are coded as poor, those that will get adopted but will offer on some improvement over the current state are coded medium, and finally, technologies that have significant potential to affect current operational energy usage are coded as high. For a technology that may be classified in more than one of these three categories based its implementation, the categories are hyphenated and used as a combination.

In the current scenario, individual energy technologies do not appear to have strong implications on lethality since the primary objective is to enhance the ability of the ground troops to remain in the field longer and reduce the load carried without losing capability. While this would have second order effects on lethality, it does not appear to be the primary goal of storage and generation technology at the individual ground troop level. The safety of the battery is projected to improve as the technology improves. Further, the conformable battery fits easily in the existing vest design while the foldable PV applications can also be easily stored in the back of the vest. This allows for a high ranking on the survivability/protection parameter. This would also imply improvements in survivability of the soldiers as they are able to be in the field for longer without risking exposure for finding fuel sources for recharging their batteries. Both technologies are used to support the ground troops in lasting longer and in enhancing their agility. Lighter batteries will allow soldiers to move faster, thereby contributing positively to their mobility; and higher energy density of conformable battery technology would also allow missions to last longer with the same weight.

Based on the length of process involved in research, prototyping, field testing, and large scale adoption of these technologies, it seems unlikely that revolutionary, wearable system-based battery technologies will be adopted in the near term, thereby limiting the capacity to cause major disruption in the energy storage and transfer process as the individual as well as fixed location level. If, however, more integrated systems get adopted at a wide scale, there can be applications in raising the lethality of troops by leveraging the increased mobility, allowing them to be even more effective in missions.

Improvements in ground vehicle propulsion are aimed largely at improving the mobility of vehicles and to make them useable for longer time periods in rugged terrains. This has important implications for the mobility of not just the vehicles, but also the troops and materials they carry. However, the survivability is generally conceived as depending on factors that most often increase energy demand, such as heavier vehicles, rather than on the underlying energy technology. Therefore, the survivability is ranked at medium-high levels. However, this hints at the complicated interdependency of energy and other factors that may not be apparent. If a material of lesser weight offers the same level of protection (i.e., armor) then the energy (in fuel) required is less. Finally, except in the case where energy is being used for directed energy weapons, the lethality of the improved technology may be altered substantively with improved fuel and propulsion options and will largely depend on the vehicles on which improved propulsion technologies are implemented; therefore, the expected rating is medium. On the other hand, the primary goal of directed energy weapons is to enhance lethality. This is also true of the mobility afforded by directed energy weapons. However, since mobility may not even be the key objective of the technology, no ranking is provided on the parameter in Table 1.

In the context of fixed bases, battery storage is likely to be one of the key contributors to future capabilities. As better storage technologies that are more power dense and therefore lighter are developed, the bases will become independent of the tyranny of distance when sourcing petroleum based fuels and the dangers thereof [39, 40]. This will ease the physical processes of setting up and moving from one location to another and the political processes of negotiating access via other nation-states, potentially resulting in an improvement of the mobility of the bases and troops and increased options for decision-makers.

Since energy storage is not expected to significantly alter the ability to generate more damage, it is ranked low on lethality. Finally, examining new concepts for nuclear fusion reactors, these have high level of mobility as they are expected to be portable. However, since the technology is nascent and significant political challenges are likely to arise, it is difficult to assess if it will be possible to disconnect and reconnect them with ease when locations and bases change.

To summarize, several developments and research projects are underway to improve the way energy is produced and consumed by the military, and are likely to have affect the long-term strategy and perhaps even alter the nature of conflict, especially with directed energy weapons. Since the weapons such as rail guns and area denial are much better at deterring aggression and reducing collateral damage, they can be used in protecting the bases and allow ground forces to change their strategy when engaging with the aggressors during conflict.

7 Programmatics

In addition to examining the different technologies, it is pertinent to analyze the programmatic aspects of energy technology research and acquisition by the US military that underlie the research and development process. This has broader implications on the institutional structure and response mechanisms and can have lasting impact on how organizations views energy and related technologies, such as how accepting they are of new technologies.

7.1 Research and Development

From an initial assessment of the programmatic aspects of energy and related technology development, it is evident that all services conduct research activities to develop technology that meets their requirements. For example, the Army Research Lab, the Office of Naval Research, the Marine Corps, the Air Force, and other agencies within each of the key services individually conduct, contract, and fund research. Independent agencies within the DOD, such as the Defense Advanced Research Projects

Agency (DARPA), also have programs and projects related to energy, as well as research and development executed by and for other parts of the US Government, such as the Department of Energy. Additionally, research conducted independently by industry (not directly funded by the DOD or other government agencies) may have dual use applications for the military, such as that being developed by Tesla and others for automotive applications. Together, these are available to form a broader portfolio of energy-related technological developments for implementation by the military. The existence of multiple research endeavors can be beneficial in that different agencies have their own priorities, while systemic differences allow for a broader range of technologies to be developed and examined. However, from the point of view of developing potential disruptions, it may be useful to look at alternative models that can lead to more fundamental changes in the way technology related to energy is considered and developed, especially in combat applications.

In the current development and acquisitions framework, a technology undergoes many steps before being adopted at a large scale for defense usage [41]. The underlying process begins with need identification that involves the decision that a new product is needed while different solutions are identified. This is followed by a "risk reduction decision" in the technology maturation and risk reduction phase. Finally, the development takes place in the engineering and manufacturing development phase and involves Capability Development Document (CDD) validation, Request for Proposal (RFP) release, and development decision. Next, low-rate initial production (LRIP) or limited deployment and operational test phases occur, where the initial base is established and limited fielding of the product takes place. After this, production can be scaled up in the production, deployment and sustainment phase where deficiencies (if any) are resolved before the large-scale deployment of this phase. There are technical as well as bureaucratic steps before any new product, or technologically-enabled capability can be adopted on a wide scale. This model allows for adequate redundancy in the program and ensures that technology has been adequately tested and deployed at small scales before a wider adoption.

Since most of the research is typically incremental, there is room for altering paths and incorporating technological changes at every step but not for making sweeping and sudden changes. The introduction of Operational Energy and the concerted move to release information and implement programs to meet the three objectives identified in the first section of this chapter provides an opportunity which has the potential to improve the approach to developing and implementing energy technology for use in military. The adoption of energy Key Performance Parameters (eKPP) [42, 43], focus on joint capabilities, and managing logistics are all measures likely to contribute significantly towards integrating the forces' ability to undertake major projects on energy technology.

Another observation regarding the military's approach to energy innovation is rooted in its historical approach to innovation at large. The Cold War was a critical determining factor in the adoption of US defense forces' policy on innovation [44]. DARPA was able to contribute significantly and the technology developed for defense use was also available for civilian projects fields (for example the internet). Innovation has also flown from the military to civilian application in the energy landscape—in areas such as commercial application of nuclear power and use of gas turbines. Renewable energy technology, battery storage, micro-grids have all been implemented in civilian usage of energy before adoption by the military. The focus of the military has been on protection and efficiency while at the same time, the pressure has been growing to reduce spending and the need to adopt technology that provides the service at the lowest cost [45]. This affects the capability to fundamentally alter or disrupt the way technology is examined since its visibility is already high before it gets adopted by the military.

7.2 Acquisitions within the Special Operations Community

In 1991, US Special Operations Command (SOCOM) established its own standalone Acquisition, Technology, and Logistics (AT&L) Center to enable tailored responses to meet its mission and rapidly respond to the needs of Special Operations Forces (SOF) [46]. The SOCOM commander has the authority to develop and acquire equipment peculiar to special operations. Major Force Program-11 provides special appropriations to the SOCOM to develop, acquire and sustain activities. Further, the small size and specific requirements of the SOCOM allow for faster acquisition processes [47].

This provides a mechanism for testing new technologies before they are transitioned for more expanded usage by all the services [48]. The Program Executive Offices under the Special Operations Research, Development, and Acquisition Center (SORDAC) Command are focused on specific needs of the SOF, but at the same time, it can leverage the smaller size of SOF to test if technologies are interoperable and deployable at a larger scale. SOCOM has the ability to more rapidly field and test several technology options. This was demonstrated in the case of acquisition of the AC-130J gunship in 2009 by SOCOM. The Joint Acquisition Task Force (JATF) was created to lead the Dragon Spear Program and report directly to the SOCOM acquisition executive. The urgent deployment acquisition process facilitated a response within 20 months, and the new technology was deployed for use in Operation Iraqi Freedom by 2011 [49]. Arguably, developing a JATF like the one created for the AC-130J program in 2009 for energy technology adoption within SOCOM could provide a useful way forward.

8 Recommendations

Several key recommendations emerge based on the preceding discussions and analyses. These can broadly be classified into research related directions for future investment and policies that affect program and acquisition policies, which are shown in Fig. 4.

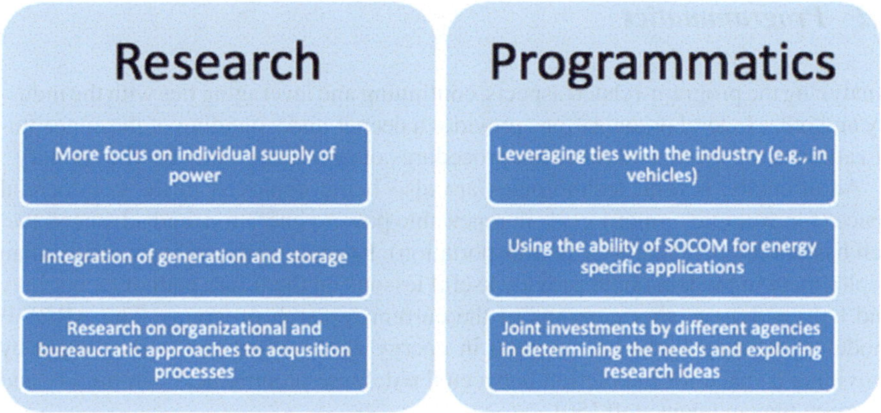

Fig. 4 Recommendations on advancing the technology application in military energy usage

8.1 Research

In terms of research, individual applications of energy technology appear to have the largest marginal gains due to their effect on mobility and survivability. As advanced energy storage systems develop, integrated power management technologies at the individual level will help provide power seamlessly for the multiple and evolving applications. This must be achieved without encumbering troops with additional loads. In fact, the load carried must be reduced in order to enhance mobility.

Furthermore, in terms of the broader applications of technology, integration of power generation with storage will become plausible as renewable sources are harnessed at all three levels—individual, vehicles, and fixed bases. In the case of individual energy production, as technology to harness kinetic energy is adopted more widely, developing light-weight, energy-dense storage technology will assume more importance. It can be argued, for these purposes, that the development of storage techniques should ideally move in tandem with each other. Finally, although the development of technology will play a key role in determining the future course of energy related developments in the military, exploring alternative processes and mechanisms for the organizational structures that affect acquisition processes for R&D and deployment in the military will also be useful. In order to achieve this, engaging with scholars and practitioners working in the fields of organizational and public management would provide useful insights.

8.2 *Programmatics*

Analyzing the program-related aspects, continuing and leveraging ties with the industry and other federal organizations provides a deeper understanding of the conceptualization, acquisition, and retention procedures of new technology for the military.

Additionally, several technologies are also being deployed in the commercial space (for example—micro-grids in renewable power generation, and advanced fuel cell based vehicle technology in transportation). Examining the military and civilian deployment in tandem could provide useful lessons on the future path of technology and help anticipate shortcomings of the current research programs. The ARPA-E model of funding high-risk research in energy applications can be leveraged by providing a space for interaction between the defense procurement community and program management staff [50].

Next, as noted earlier, SOCOM has a stand-alone technology development and acquisition process. This process can be used for developing technologies that cater to specific requirements for SOCOM can be deployed more quickly, at a smaller scale before the usage is expanded to all services in the military.

Finally, developing a task force that proactively explores evolving technologies in the field of energy in both commercial and non-commercial applications and works to tailor them for military usage can be useful. This would not only help in changing the role of energy from a logistics oriented factor, but it would also help reduce lag time between ideation and wide scale deployment of new technologies and reducing any surprises or disruptions.

9 Conclusions

Incorporating better and more efficient forms of energy consumption provides advantage to the military in various forms of conflict. It allows troops on ground to conduct operations for longer periods of time while reducing the loads they need to carry. Battery powered vehicles can move more swiftly and quietly in the battlefield while improving their range and firepower through energy weapons. Facilities and bases can effectively reduce, if not eliminate, the tyranny of energy transportation logistics. This helps mitigate vulnerabilities to both infrastructure and human life. However, these technologies do not fundamentally alter the way of producing and consuming energy. Therefore, while the developments noted in the field are significant, they may not necessarily meet the criteria of being ‘disruptive.’ Changes are likely to be incremental and not fundamental. This is partly a result of the nature of the energy value chain itself where technology usage cannot be phased in till it is fully developed and duly tested. However, the programmatic aspects of technology development in the military also play a significant—and perhaps under-recognized—role in determining the types of projects that various research groups embark on and the applicability of the technologies developed therein. This also has broader implications on the approach to acquiring and using energy technologies by the military.

Energy is a key logistical input in the functioning of the military. The current systems are based on technology and policy related decisions that have accumulated over many decades. The legacy creates a systemic path dependence where all new energy applications must be altered enough to be to be compatible and inter-operable with the existing system without changing the larger support/infrastructure mechanism. This is reflected in the type of energy used, the technology and equipment used for harnessing the technology, and certain types of energy availability and technology in both civilian and military usage. For example, in the case of batteries, Lithium Ion based batteries have been in use for a long period of time. Further, any new design needs to conform with the suits currently worn by service members. Similarly, for vehicles, the fuel needs to be of a form that can be used in the current fleet of ground vehicles, marine fleet, and aircrafts. This adds to the challenge of technology design and implementation, and arguably, limits the ability of any new technology concept to significantly disrupt the way energy is produced and consumed, especially for combat operations.

These limitations notwithstanding, there are potential gains to be made in specific areas of research as well as program based applications. Specific technologies that help reduce the load for the ground troops, and help integrate energy production and storage have the potential to provide substantial gains. On the organizational front, focusing on energy and leveraging the ability of innovative organizations and processes to rapidly develop, acquire, and field technology options will help in increasing the speed at which technology gets adopted.

In short, only the logical fallacy that underlies the "slippery slope" argument indicates that NLW will cause more harm than good to our world. Thus, another roadblock to the development and deployment of NLW has been moved aside.

References

1. Bohnwagner A (2015) Considering energy logistics in force development. Office of the Assistant Secretary of Defense for Sustainment, 16 Sept 2015. https://www.acq.osd.mil/eie/Downloads/OE/Considering%20Energy%20Logistics%20in%20Force%20Development%2020150916.pdf
2. Sarewitz D et al (2012) Energy innovation at the Department of Defense: assessing the opportunities. Consortium for Science, Policy and Outcomes at Arizona State University, Mar 2012. http://cspo.org/wp-content/uploads/2014/11/Energy-Innovation-at-DOD.pdf
3. National Security Strategy of the United States of America, White House, Dec 2017. https://www.whitehouse.gov/wp-content/uploads/2017/12/NSS-Final-12-18-2017-0905.pdf
4. Operational Energy Strategy. Office of the Assistant Secretary of Defense for Energy, Installations and Environment, US Department of Defense 2016. https://www.acq.osd.mil/eie/Downloads/OE/2016%20DoD%20Operational%20Energy%20Strategy%20WEBc.pdf
5. Operational Energy. Office of the Assistant Secretary of Defense for Energy, Installations, and Environment. http://www.acq.osd.mil/eie/OE/OE_index.html

6. Operational Energy Strategy (2016) Office of the Assistant Secretary of Defense for Energy, Installations and Environment, US Department of Defense. https://www.acq.osd.mil/eie/Downloads/OE/2016%20DoD%20Operational%20Energy%20Strategy%20WEBc.pdf
7. International Science and Technology Engagement Strategy (2014) US Department of Defense. Strategic Guidance. https://defenseinnovationmarketplace.dtic.mil/wp-content/uploads/2018/02/2014InternationalStrategyASDRE.pdf
8. Army Equipment Modernization Strategy: Equipping the Total Force to Win in a Complex World. G-8 Future Force Division, Office of the Chief of Staff, Department of the Army, 27 Mar 2015. http://www.g8.army.mil/pdf/AEMS_31MAR15.pdf
9. Shaffer E (2014) U.S. military energy and power S&T. Innovative Solutions for Military Applications 12 Nov 2014. https://www.iesma.info/download/125/8
10. Army Equipment Program in Support of the President's Budget (2015) G-8 Future Force Division, Office of the Deputy Chief of Staff, Department of the Army, May 2014. http://www.g8.army.mil/pdf/army_equipment_program2015.pdf
11. ADP 3-0: Unified Land Operations. Headquarters, Department of the Army, 10 Oct 2011. https://www.army.mil/e2/rv5_downloads/info/references/ADP_3-0_ULO_Oct_2011_APD.pdf
12. Joint Engineer Operations. Joint Publication 3-34, Joint Chiefs of Staff, 6 Jan 2016. https://www.jcs.mil/Portals/36/Documents/Doctrine/pubs/jp3_34.pdf
13. Air Mobility Operations. Joint Publication 3-17, Joint Chiefs of Staff, 5 Feb 2019. https://www.jcs.mil/Portals/36/Documents/Doctrine/pubs/jp3_17.pdf
14. Fiscal Year 2015: Operational Energy Annual Report. Office of the Under Secretary of Defense for Acquisition, Technology, and Logistics, Department of Defense, Aug 2016. http://www.acq.osd.mil/eie/Downloads/OE/FY15%20OE%20Annual%20Report.pdf
15. Hicks S (2017) Powering the department of defense: initiatives to increase resiliency and energy security: fact sheet. American Security Project, Sept 2017. https://www.americansecurityproject.org/wp-content/uploads/2017/09/Ref-0204-Powering-the-DoD.pdf
16. Fiscal Year 2017: Operational Energy Annual Report. Office of the Under Secretary of Defense for Acquisition and Sustainment, Department of Defense (2018). https://www.acq.osd.mil/eie/Downloads/OE/FY17%20OE%20Annual%20Report.pdf
17. Ground Renewable Expeditionary Energy Network System (GREENS). UEC Electronics (2016). http://www.uec-electronics.com/?s=ground+renewable+expeditionary+energy+network+system+%28greens%29
18. Lockheed Martin (2014) Compact Fusion Research & Development. Lockheed Martin, 15 Oct 2014. https://www.youtube.com/watch?v=UlYClniDFkM
19. Cohen A (2018) Will Lockheed martin change the world with its new fusion reactor? Forbes, 1 Aug 2018. https://www.forbes.com/sites/arielcohen/2018/08/01/will-lockheed-martin-change-the-world-with-its-new-fusion-reactor/#432592b94c49
20. Langston J (2015) UW researchers scaling up fusion hopes with DOE grant. UW News, 02 June 2015. http://www.washington.edu/news/2015/06/02/uw-researchers-scaling-up-fusion-hopes-with-doe-grant/
21. King J (2017) The overweight infantryman. Modern War Institute at West Point, 10 Jan 2017. https://mwi.usma.edu/the-overweight-infantryman/
22. Hambling D (2018) The overloaded soldier: why U.S. infantry now carry more weight than ever. Popular Mechanics, 26 Dec 2018. https://www.popularmechanics.com/military/research/a25644619/soldier-weight/
23. Murphy P (2011) Weight of war: soldiers' heavy gear packs on pain. National Public Radio. https://www.npr.org/2011/04/10/134421473/weight-of-war-soldiers-heavy-gear-packs-on-pain
24. Vergun D (2018) Army researchers hope to lighten soldiers' battery load. US Army News Service, 4 Sept 2018. https://www.army.mil/article/210673/army_researchers_hope_to_lighten_soldiers_battery_load
25. Wang JL. Piezotronics for smart systems. Research Highlights. http://www.nanoscience.gatech.edu/research/

26. Unburdening the Soldier through innovations in battery, power component technology (2011) United States Army Research Laboratory, 15 Mar 2011. http://www.arl.army.mil/www/?article=564
27. Soldier Wearable Integrated Power Equipment System (SWIPES) (2019) US army combat capabilities development command C5ISR center, last modified, 26 Feb 2019. https://www.cerdec.army.mil/news_and_media/Soldier_Wearable_Integrated_Power_Equipment_System/
28. Thampan T (2015) Soldier wearable power system. Communications-Electronics Research, Development and Engineering Center, Army Materiel Command, United States Army. https://science.dodlive.mil/files/2015/10/DOD_Thampan.pdf
29. Soldier Wearable Integrated Power Equipment System (SWIPES) (2019) US Army Combat Capabilities Development Command C5ISR Center, 26 Feb 2019. https://www.cerdec.army.mil/news_and_media/Soldier_Wearable_Integrated_Power_Equipment_System/
30. Stennis JC (2016) The great green fleet explained. Strike Group Public Affairs, US Navy, 27 June 2016. https://www.navy.mil/submit/display.asp?story_id=95398
31. Fiscal Year 2015: Operational Energy Annual Report. Office of the Under Secretary of Defense for Acquisition and Sustainment, Department of Defense, Aug 2016. http://www.acq.osd.mil/eie/Downloads/OE/FY15%20OE%20Annual%20Report.pdf
32. Vergun D (2017) Army showcases stealthy, hydrogen fuel cell vehicle. US Army News Service, 27 Jan 2017. https://www.army.mil/article/181342/army_showcases_stealthy_hydrogen_fuel_cell_vehicle
33. Law D (2016) Next-generation non-lethal technologies. The International Society for Optics and Photonics, 14 Sept 2016. http://spie.org/newsroom/6484-next-generation-non-lethal-technologies
34. Non-Lethal Weapons (NLW) in NATO and National Concepts. Non-Lethal Weapons Program, US Department of Defense, 11 Aug 2017. https://jnlwp.defense.gov/Press-Room/In-The-News/Article/1300484/non-lethal-weapons-nlw-in-nato-and-national-concepts/
35. Directed Energy: Sound and Light Technology. Non-Lethal Weapons Program, US Department of Defense, 13 July 2018. https://jnlwp.defense.gov/Press-Room/Fact-Sheets/Article-View-Fact-sheets/Article/1574157/directed-energy-sound-and-light-technology/
36. Active Denial Technology Fact Sheet (2016) Non-Lethal Weapons Program, US Department of Defense, 11 May 2016. https://jnlwp.defense.gov/Press-Room/Fact-Sheets/Article-View-Fact-sheets/Article/577989/active-denial-technology/
37. Beidel E (2014) Bad news for the bad guys: laser weapon being readied for marine vehicles. Office of Naval Research, 11 June 2014. https://www.onr.navy.mil/en/Media-Center/Press-Releases/2014/Ground-Based-Air-Defense-Laser
38. Eckstein M (2016) Marines to pair laser weapon with stinger missile for mobile ground unit protection. USNI News, 23 June 2016. https://news.usni.org/2016/06/23/walsh-directed-energy
39. Helman C (2009) For U.S. military, more oil means more death. Forbes. https://www.forbes.com/2009/11/12/fuel-military-afghanistan-iraq-business-energy-military.html#3cedc5804562
40. Bohnwagner A. Considering energy logistics in force development. Office of the Assistant Secretary of Defense for Sustainment, Sept 15. https://www.acq.osd.mil/eie/Downloads/OE/Considering%20Energy%20Logistics%20in%20Force%20Development%2020150916.pdf
41. Generic Acquisition Process (Pre-Tailoring) Defense Acquisition University, 2014. http://acqnotes.com/wp-content/uploads/2014/09/Generic-Acquisition-Process-Pre-Tailoring-Chart.pdf
42. Duncan Hunter National Defense Authorization Act for Fiscal year 2009. Public Law 110-417, 110th Congress 14 Oct 2008. https://www.govinfo.gov/content/pkg/PLAW-110publ417/pdf/PLAW-110publ417.pdf
43. Bohnwagner A (2013) An overview of the energy key performance parameter (KPP). Office of the Under Secretary of Defense for Acquisition, Technology and Acquisition, Department of Defense, 09 Dec 2013. https://www.acq.osd.mil/EIE/Downloads/OE/Energy%20KPP_12-09-13.pdf
44. Energy Innovation at the Department of Defense: Assessing the Opportunities. Consortium for Science, Policy and Outcomes, Mar 2012. http://cspo.org/wp-content/uploads/2014/11/Energy-Innovation-at-DOD.pdf

45. Strakos JK, Quintanilla JA, Huscroft JR (2016) Department of defense energy policy and research: a framework to support strategy. Energy Policy 92:83–91
46. SOF AT&L. United States Special Operations Command. http://www.socom.mil/SOF-ATL/
47. Schwartz M, Purdy JA (2018) United States Special Operations Command Acquisition Authorities. Report R45252 Congressional Research Service, 09 July 2018. https://fas.org/sgp/crs/natsec/R45252.pdf
48. Guerts J (2014) Special Operations Research, Development, & Acquisition Center. http://www.afceaboston.com/documents/documents-briefings/SOCOM%20AFCEA%20Brief%202014.pdf
49. Breede D (2014) Soaring with AC-130J. Defense Acquisition University. https://www.dau.mil/library/defense-atl/p/Defense-ATandL---July-August-2014
50. Bonvillian WB, Van Atta R (2012) The energy technology challenge: comparing the DARPA and ARPA-E models. In: Energy innovation at the department of defense: assessing the opportunities, consortium for science, policy and outcomes. https://cspo.org/legacy/library/1203281138F90409638BA_lib_EnergyInnovation.pdf

Meta-materials: Threat to the Global Status Quo?

Margaret E. Kosal and J. Wes Stayton

Abstract This chapter investigates the disruptive potential of metamaterials, synthetic nano-structured materials, which possess special physical properties that enable the user to be invisible and to evade detection. The security applications of metamaterials are to camouflage personnel, vehicles, ships, or planes from some portion of the infrared spectrum. Metamaterials have a high refractive index meaning that light 'flows around' the material rather than reflecting off. Successful implementation of metamaterial adaptive camouflage (MMAC) would be a paradigm shift in camouflage and anti-detection technology that could cause significant disruptions to conflict dynamics. This chapter investigates the role of metamaterials by placing it in the broader historical context of detection and anti-detection technology. By tracing the co-development of both sides of the detection arms race, a theoretical framework can be constructed. This analysis pays special attention to radar and stealth aircraft in the post WWII era. It explores concepts like deterrence, the offense defense balance, and the security dilemma. The conclusion is that anti-detection, or camouflage, technologies are generally detrimental to the international peace and detection technologies are beneficial. Metamaterials are a threat to the global status quo and are therefore a threat to the power of global and regional hegemons. Revisionist actors, primarily non-state actors, likely will benefit disproportionately from acquiring a MMAC capability but will struggle to do so due to the technical challenge of advanced R&D, particularly in the near and mid-term. The implication is that status quo powers—whom are likely to be the first to develop a viable capability—must emphasize parallel development of countermeasures and limit the negative potential of the technology's proliferation.

M. E. Kosal (✉) · J. W. Stayton
Georgia Institute of Technology, Atlanta, GA, Georgia
e-mail: margaret.kosal@inta.gatech.edu

J. W. Stayton
e-mail: jack.wesley.stayton@gmail.com

M. E. Kosal (ed.), *Disruptive and Game Changing Technologies in Modern Warfare*, Advanced Sciences and Technologies for Security Applications,
https://doi.org/10.1007/978-3-030-28342-1_8

1 Introduction

In November 2016, Lieutenant General HR McMaster, then-Deputy Commanding General for Futures and Director of the Army Capability Integration Center (ARCIC) at the US Army Training and Doctrine Command (TRADOC), invoked the concept of invisible tanks at meeting on the Army Ground Combat Platforms at the AUSA Institute of Land Warfare [1]. One of the subsequent speakers noted the need for new capabilities that shift the approach to survivability from protection via mass (which is limiting) to capabilities for active defense and active camouflage such as via meta-materials [2]. Technical challenges of needing to expand the size and scale of objects being cloaked and the width of spectrum were presented, as well as the need to make such a capability tunable along the electro-magnetic spectrum. Operational and strategic requirements in near-peer competitor or dense urban (aka megacities) operating scenarios were highlighted, particularly in the context of the end of the 'era' of physical mass providing protection fast approaching.

Invisibility cloaks have been a regular staple of science fiction from Jules Verne's *The Invisible Man*, which features a yellow fluid created by a Prussian scientist that enables people to become invisible, to Star Trek's Romulan cloaking device, which camouflages an entire space craft. What does it mean for international security when states or non-state actors have access to invisibility cloaks?

While this chapter focuses on the implications of one specific type of technology, meta-materials, the underlying question of how technological advancements impact the state of international peace is a much broader one. According to Offense-Defense Theory, technologies that favor attack reduce the likelihood of maintaining peace, while those bolstering defenses increase that likelihood. This chapter argues that technologies which increase the ability of a state to detect the actions of another favor the defense and are therefore stabilizing while those that decrease the ability to detect favor the offense and are therefore destabilizing.

This chapter attempts to explore, and where possible answer, two questions, one theoretical and another practical. First, the proposition that detection-enhancing technologies favor stability while detection avoidance technologies, such as active camouflage, do the opposite will be examined. More practically, this paper also explores the future impact of metamaterial adaptive camouflage, a fledgling technology with the potential to disrupt the entire detection and intelligence paradigm. Beyond that November 2016 discussion cited at the start of this section, little has been explored regarding meta-materials based active camouflage in the context of national security and international affairs [3].

In order to predict the effects of metamaterial adaptive camouflage, a history of detection and anti-detection (or evasion) technologies and their impacts on the stability of the international system is presented. This strategic security analysis is done through the lens of various theories of conflict, most prominently Offense-Defense Theory.

One of the most concepts most useful for thinking about evasion and detection technologies are Cold War nuclear deterrence and the 'Security Dilemma.' As with other technologies, cloaking technologies—including metamaterial adaptive camouflage—cannot be assigned a purely offensive or defensive purpose because they do not exist in a vacuum. The use of all technologies is dependent upon both the technological ecosystem and international political climate in which they exist. In general, the proliferation of information about the intentions of other actors in the international system is good for international peace. However, due to differences in perception among states, we can never accurately claim to know the effect of any individual technology on the stability of the system.

Intuitively, metamaterials make detection more difficult which reduces information, making the technology destabilizing to the international system. However, some historical developments of stealth suggest this is not always the reality. This investigation on the effect of metamaterials on stability is inconclusive. There are also doubts about whether metamaterial cloaking is a revolutionary technology, one might hypothesize it is merely a continuation of the decades old stealth arms race. Its revolutionary impact will depend on the relative development of cloaking and anti-cloaking capabilities. A lesson of history on the development of stealth technologies is that stealth innovations and their respective countermeasures often develop concurrently. When that doesn't happen, as with stealth aircraft capabilities in the late 20th Century, one side may realize a distinct asymmetric advantage. Additionally, how the diffusion of metamaterial technology to non-state actors will affect the stability of the international system is not known. If terrorist organizations or other non-state actors were to acquire stealth capabilities, it also could be massively disruptive.

2 Metamaterials

Adaptive camouflage, or active camouflage, is a technology which allows the user to totally conceal itself from plain sight. Other proposed variants of adaptive camouflage include cloaking from different parts of the infrared, visible, and ultraviolet spectrum. Research and development in these types of nano-structured materials, also known as meta-materials, for active camouflage is ongoing in academic and in industry [4]. While meta-materials have potential and demonstrated applications for a variety of technologies where electromagnetic radiation is used, such as quantum computing, optical data processing, superconductivity [5]; this chapter focuses on applications for cloaking, stealth, and adaptive camouflage. It is truly an emerging technology.

Structural nanomaterials have already been incorporated into security applications such as up-armored Humvees to protect against improvised explosive devices (IEDs) and personal protective gear from chemical and biological agents, functional nanomaterials may afford even greater capabilities [6]. Meta-materials are synthetic structures that bend light obscuring an object from view [7–9]. Today most meta-materials are limited to fairly narrow regions of the visible or infrared spectrum; advances in the technology will widen the effective part of the light spectrum cloaked [10, 11]. Biologically-inspired or biomimetic offer routes to functional materials that offer additional routes to new capabilities [12–14].

What is adaptive camouflage? Adaptive camouflage development is patterned after biological cloaking systems used by reptiles, amphibians, and fish [15]. The goal of active or adaptive camouflage is to make a person, vehicle, or weapon invisible to enemy much like an animal conceals itself from a predator or prey. Invisibility is achieved by altering either color or luminescence. Scientists believe the best chance for humans to replicate the cloaking capabilities of animals is the development of metamaterials—synthetic materials exhibiting unique properties with respect to angle of refraction of light (or other) waves [16]. Metamaterials, essentially, have the ability to bend electromagnetic radiation—visible light, radar, infrared—giving the illusion that the material isn't present. In technical terms, this is referred to as negative index of refraction or overly simplified bending light so that an object can't be seen [17]. There are certainly challenges impeding the development of a true cloak of invisibility, but much of the informed speculation from the scientific community is cautiously optimistic about the future of adaptive camouflage technology. It is not possible to assess with perfect fidelity the current state of adaptive camouflage technology due to the uncertainty in research and the sensitive nature of national security technology development. However, some of the potential of adaptive camouflage technology may be inferred by examining its scientific foundations.

While the idea of light-bending optics dates back to the start of the 20th Century, the theoretical physics of such was derived during the high of the Cold War [18]. The scientific breakthrough came in 2006, when Duke University physicist David Smith and team created a microwave bending metamaterial [19]. Smith's cloak used copper spring resonators and only worked in two dimensions. The concept was advanced by replacing the copper with gold and layering them over a synthetic silk that only interacts with a restricted region of the electromagnetic spectrum (terahertz waves). Since then successive researchers have expanded to additional materials and additional wavelengths and from 2D to 3D camouflage capabilities [20–27].

These synthetic materials derive their visual characteristics from their chemical compositions. These materials have a negative refraction index resulting from the materials' variable permittivity and permeability. Ostensibly, the material can rearrange its structure to accommodate varying levels of interaction with the electromagnetic spectrum. In essence, manufactured meta-materials woven into a surface can be configured in such a way as to deny interaction with subsets of the electromagnetic spectrum—including light. The surface does not reflect light, rather light flows around the surface like water around a stone in a stream.

The development of metamaterials is a rapidly developing field. In some cases, such as chalcogenide glasses, meta-materials have been commercially available for decades, most commonly in rewritable optical disk drives. Prototypes are in development including, such as the ADAPTIV Cloak of Invisibility from BAE systems, which provides the user with the ability to cloak a vehicle with a honeycomb cover that can adjust the projected appearance and temperature of the vehicle to match the surroundings or mimic another vehicle [28]. The company claims the technology could be used to conceal anything from trucks to helicopters and even buildings. ADAPTIV conceals the vehicle from IR detection but does not offer plain sight disguise.

Beyond university and government laboratories and outside private industrial entities, which are busy developing adaptive camouflage technology, do-it-yourself (DIY) technologists and artists are also attempting to use existing technology to fashion their own cloaks of invisibility [29].

Challenges for operational usage remain that need to be overcome. The materials are expensive to create, and there are scalability issues due to limitations in the fabrication process of a large metamaterial surface, fabrication is done on the scale of micro and nano-meters. There is also the challenge of broadening the range of angles at which invisibility is achieved. Like all advanced materials, challenges of the durability of materials in operational conditions exist.

As more research is done on metamaterials, and specifically the mass production of metamaterial cloaking surfaces, adaptive camouflage will become a more practical technology. In addition to security-related purposes, there currently are R&D efforts directed to developing metamaterials for medical applications [30]. A shift toward camouflage specific R&D will lead to quicker development of metamaterial adaptive camouflage. The basic science behind MMAC is progressing relatively quickly. In order to be effective in a realistic environment, the applied research stage will have to reduce manufacturing costs and address the environmental challenges facing MMAC.

In can be ignored that conflict occurs in generally harsh and non-permissive environments where equipment is subject to the prolonged extreme stress of everyday use. Concepts of operations (ConOps) will have to be developed, tested, and incorporated into doctrine and training. Like other technologies, the cost of manufacture and maintenance has to fall substantially or the strategic value has to be determined to be high enough that cost is not the critical factor. The durability of the material has to increase. In desert conditions, sand, dust, heat, and other contaminants must not significantly degrade a capability. In tropical conditions, humidity will have to be addressed. Research that addressees challenges such as is ongoing [31, 32], and, in the fast-moving world of scientific discovery, it is difficult to stay at the cutting or bleeding edge. In conventional environments like the air and sea there are also challenges. Salt water is notoriously demanding. Any attempt to cloak ships or small boats with metamaterials would require special research and development into salt water compatibility. Aircraft may have similar or their own limitations due to the extreme conditions of the troposphere and fluid dynamics at velocity.

These environmental challenges suggest two initial possible engineering or material-focused routes for the development of a practical metamaterial adaptive camouflage. The manufacture of robust materials that can withstand harsh conditions over extended durations, or low cost materials that can be quickly and affordably applied and re-applied e.g., paint-on camouflage. Like all emerging technologies, the engineering and development questions of operationalizing the materials outside of lab environmental conditions remain to be addressed.

Stealth does have a quality all of its own. The skins of stealth aircraft are comparatively delicate to non-stealth exteriors. Conceptually stealth enables aircraft to evade detection, thereby reducing the need for protection in the form of exterior shielding. A conceptual shift in designing ground vehicles that move like stealth aircraft is less likely that resolving the engineering challenges mentioned in the previous paragraph.

3 The International System

In order to assess the impact of emerging adaptive camouflage-related technologies, characterization of the international system and basic assumptions about the nature of conflict are needed. Underpinning this analysis are three basic principles: the international system is anarchic, the intentions of a state cannot be assumed, and the state is motivated by survival. Consequently, states that rationally strive to ensure their own survival will seek to do so through the accumulation of power. There is a debate among scholars over what constitutes power [33]. Is it only material capabilities and personnel capacity (manpower)? Or does it also include intangible like membership in international organizations and cultural and political influence? For our purposes, this works will assume power includes both the physical capabilities and capacity (to include population) of the state and, to a lesser extent, its ability to influence international arms control regimes and laws of armed conflict, such as the ones that will likely influence the development of camouflage capabilities.

In such an international system, a 'Security Dilemma' exists [34], which has the potential to increase the likelihood of conflict. Under the security dilemma, states amass capabilities in order to maximize their own security. Simultaneously they also unintentionally threaten the relative security of other states and compel those other states to hedge to ensure their own security. While the development of a weapons system could ostensibly be for defensive purposes according to the state developing a capability, their neighbor may not perceive their intentions as peaceful and instead declare the new weapon to be offensive.

No matter how innocuous a new capability may seem, there are few exceptions to the assumption that intentions cannot be presumed. Therefore, uncertainty is created and states are likely to respond. For example, a missile defense system can be interpreted as tactically offensive. The defense system could theoretically be reconfigured in the field to fire on targets other than incoming missiles, such as personnel or material. Essentially defensive technologies can also be strategically offensive. Missile defense systems create a second strike liability for the adversary. In nuclear combat, deterrence is achieved through the guarantee of massive retaliation which is only possible with a viable second strike capability—the ability of the target state to return a nuclear volley following an attack, lowering the benefits and raising the costs of initiating a nuclear war. Missile defense systems provide the capability for the initial attacker to successfully defend against a weakened second strike. Without the threat of massive retaliation, the cost/benefit of a nuclear first strike is reversed, without the threat of second strike vulnerability the only way to survive nuclear combat is to initiate the conflict in order to defend again a weakened second strike rather than a full-fledged first strike.

Under uncertainty, states are compelled to hedge—that is to assume that an arms buildup is offensive in nature necessitating a responding buildup either at or above par with the initial buildup. Consequently, the subsequent buildup inspires a response to the response. The cycle can continue indefinitely and spiral into a paranoia driven arms race. This is why the concept is called a spiral model.

From a pragmatic perspective, the world's super powers and regional powers are not in such an arms race at all times, so there must be some intervening factors. Treaties, membership in international organizations, and confidence building measures, which include inspections and verification regimes, can all serve to alleviate the threat of or reduce the security dilemma. Each of these factors essentially provide information that, to an extent, reveal the intentions of rivals and reduces uncertainty. Merely the act of participating in international organizations and holding a dialogue signal a willingness to coexist peacefully. Offensive Realists, however, doubt the actual effectiveness of regimes and institutions in mitigating the security dilemma [35], whereas Defensive Realists, particularly Robert Jervis, with his Offense-Defense model of conflict and cooperation, provides a compelling model for understanding both the contextual and technological factors determining the relative stability of the international system [36]. That ideas underpin much of the analysis in this chapter.

4 Modelling Stability

What is stability? It can be assessed as the lack of great power conflict and the acceptance of the status quo of the international system. Accepting the status quo means the state is unwilling to initiate conflict in order to affect either the global or regional balance of power. States may still seek to increase their absolute and relative power, but they do so with alternative means to war. Jervis' Offense-Defense model is useful for understanding the conditions under which conflict and cooperation are likely and to probe how new technologies might affect stability. In order to answer the question about the potential revolutionary effects of metamaterial adaptive camouflage on conflict dynamics, this chapter first describes the nature of the Offense-Defense balance and then attempts to integrate technology into the model in order to predict the effects of metamaterial cloaking.

A summary of Jervis' model describing four strategic environments resulting from different types of technologies is presented in Fig. 1. "Posture" refers to the transparency of intentions by states in the international system, and "Advantage" is the relative power of offensive and defensive technologies.

If this concept was to be expressed as a mathematical equation, it would say the probability of conflict is a function of information proliferation and the ratio of offensive capabilities to defensive capabilities. More complete information on the intentions of rivals allows both sides to manage a spiraling arms race. Awareness of aggression allows for coalition building and diplomatic action in order to preemptively quell belligerence. An advantage of defensive technology over offensive technology lowers the cost/benefit equation of the attacker and, in the words of Clausewitz, "tame the elementary impetuosity of War." [37]. Offensive ascendency, conversely, creates a sense of urgency for states to develop greater offensive capabilities and seek out alliances, further increasing tensions.

	Posture Distinguishable	Posture Indistinguishable
Offensive Advantage	Offensive capabilities are ascendant, making preemptive attack a tempting strategy. However, transparent intentions and information proliferation lower the incentives for surprise attack and are conducive to diplomatic dialogue.	The intentions of peers are unknown and offensive capabilities are ascendant in this world. The security dilemma is strong due to a saturated threat environment and incomplete information about the intentions of competitors.
Defensive Advantage	This is the most stable scenario. Defensive capabilities are ascendant and intentions are well known. Both factors mitigate the security dilemma and lower the probability for conflict initiation.	Paranoia still contributes to the tense and distrusting international environment. However, the advantage of defensive technologies over offensive technologies mitigates some mistrust.

Fig. 1 Jervis' Offensive Defensive model of possible strategic environments from different technologies [36]

Offensive Realists like Mearsheimer claim that it is impossible to distinguish between offensive and defensive weapons. No weapon has either a purely offensive or purely defensive purpose. Like the missile defense system discussed earlier, states can never have enough information to assume a system cannot or will not be used in an aggressive capacity. The spiral model of conflict is not ephemeral or conditional, rather it is an unavoidable characteristic of the anarchical international system. Instead of intentions determining the propensity for conflict, only raw capabilities influence the decision to cooperate or attack. In order for the international system to be stable, status quo powers must discourage revisionist powers from initiating conflict by decisively signaling their intention to retaliate to such a degree that the cost of aggression outweighs the benefits.

Another model in international security is the "Deterrence Model," which centers on ability to clearly and credibly communicate intentions [38]. Communicating a deterrent intention is difficult because all states signal their resolve to fiercely defend their interests. A credible deterrent can only be achieved by tangibly demonstrating the capability to inflict unacceptable levels of cost on potential challenger. A nuclear power can articulate a policy of massive retaliation in the event of a nuclear strike, making the costs of aggression incredibly high. Mutually Assured Destruction, an implicitly agreed upon doctrine between the US and USSR during the Cold War, was dependent upon the guarantee of a massive retaliatory strike. Counterintuitively, the possession of a nuclear arsenal with the ability to devastate each other actually increased the security of both the US and USSR. The essential element of deterrence was the promise of a second strike. The incentive is therefore to ensure your own second strike capability while also avoiding threatening the second strike capability of the other state. If a state cannot ensure requisite retaliation, then it cannot deter aggression and its strategic calculus is altered to where the incentive to undertake a preemptive strike is now an undesirable, yet viable, option. The Anti-Ballistic Missile Treaty—signed by the US and USSR in 1972—limited the development of nuclear counter measures in order to ensure the second strike viability of both parties [39].

Which model better describes the reality of conflict and cooperation, the Spiral Model or the Deterrence Model? Do intentions matter or do raw capabilities determine the relative stability of the international system. This chapter takes both models into account when evaluating the effect of metamaterial cloaking on the international peace. The next section will explore the historical relationship between stealth technology and the relative stability of the international system and use the lessons learned to predict the effect of metamaterials.

5 History and Strategic Importance of Stealth

At the start of the 21st Century, the Defense Science Board (DSB) identified stealth as one of four "strategic vectors" that gave the US technological advantage over adversaries in the Cold War [40]. Stealth was cited specifically as increasing the "element of surprise" over the Soviet Union, particularly in context of their substantial investments in air defense capabilities. Along with stealth, the other "technologies" (i.e., capabilities) were precision, speed, and tactical intelligence, surveillance, reconnaissance (ISR).

The modern origins of stealth date back to the 17th century and the basic science of Maxwell and Hertz. James Clerk Maxwell published *A Dynamical Theory of Electromagnetic Field* in 1865 where he provided the foundation of the study of electric and magnetic waves [41]. In 1887, Heinrich Hertz proved the existence of electromagnetic waves and set the stage for the development of radio communication—beginning the wireless era. Hertz showed that electromagnetic radio waves reflected off of metallic surfaces—such as vehicles—and could be used to detect metal objects [42]. Christian Hulsmeyer conceived of the forerunner to radar detection by employing radio waves in his Telemobiloscope design [43]. Hulsmeyer's invention was outfitted on merchant ships in order to detect nearby vessels in order to prevent collisions. Although Hulsmeyer's Telemobiloscope was ultimately a commercial failure, it served as inspiration to a generation of inventors working in the electromagnetic detection space.

Rapid advances in electromagnetic detection were made in the interwar period in Europe and the United States [44]. The US, USSR, UK, France, Germany, Italy, Japan, and the Netherlands made independent advances in what would be radar technology [45]. The US Naval Research Laboratory conducted pioneering basic scientific research along with experiments on radio signal tracking of moving ships and aircraft in the 1920s and 1930s that were critical for the application of radar as a military capability [46].

The British Meteorological Office, led by Robert Watson-Watt, did innovative work on the early detection lightning storms using long wave radio. In the run-up to WWII, Watson-Watt investigated the capability of short-wave radio to detect aircraft at the behest of Air Ministry Committee for the Scientific Survey of Air Defense [47]. The War Office was concerned with the vulnerability of British east coast cities to attacks from European mainland airbases which were only 20 min' flight time from London. Before radar detection, the only way to ensure air defense were constant air patrols which were prohibitively expensive. In response, the Air Ministry developed Chain Home, a rudimentary network of coastal early warning radar stations. Chain Home was the decisive factor in the Battle of Britain and tipped the air dominance advantage in favor of the RAF for remainder of the WWII air campaign.

In the First World War, Germany was able to dominate British airspace with zeppelins unthreatened by anti-air guns. Many believed World War II would be even more unbalanced in favor of air attack due to major advances in aircraft technology and payload delivery systems, hence the saying "the bomber will always get through." [48]. However, with the development of radar and especially Chain Home, the air campaign proved to be much more competitive than anticipated. The Germans also possessed an air detection capability with the Freya system [49]. The Freya was a much more advanced system, but ultimately less significant due to the evolving nature of the war. Although the Freya was smaller, more efficient, and less prone to errors, its impact on the war was limited by the timing of the war. Chain Home was a novel system when introduced in 1939 and a critical element of the successful defense of British territorial integrity in the Battle of Britain. When the Allies turned the momentum of the war and took the offensive to continental Europe, radar technology had thoroughly exploded. Countermeasures such as chaff, and tactics like single stream bomber saturation swung the balance in favor of the offense.

The World War II air campaign demonstrated the importance of detection to both offensive and defensive operations. Post-War R&D focused on developing sophisticated radar as well as stealth penetration aircraft and eventually missiles. With the advent of nuclear weapons, the acceptable threshold of infiltration dropped considerably. The nature of a great powers air campaign had changed from low sophistication and high volume to high sophistication and low volume. The destructive density of payloads had risen exponentially, and with it, the vulnerability of ground targets. Epic air battles like The Battle of Britain were no longer the frame of reference for strategists. Although the goals of the air campaign remained the same—delivering effects to ground targets by dominating air space—the tactics for achieving dominance had changed. Instead of heavy, imprecise bombardment at close range, the focus of the new air campaign was and is avoiding detection in order to deliver a carefully selected few highly precise, highly destructive weapons at critical command and control targets.

The 1950's were a critical juncture for the development of both stealth capabilities and doctrine. The United States and allied powers heavily emphasized detection capabilities and invested heavily in radar and stealth aircraft. The first Doppler radar systems—much more sophisticated versions of the earlier Chain Home systems—introduced in the 1950's, was a revolution in air defense due to its increased precision and identification capabilities [50]. Concurrently, the Central Intelligence Agency was developing its counter to Doppler radar, the U2 spy plane. The U2 was the result of an Air Force request for proposal for an aircraft which could fly above 70,000 feet, beyond the range of Soviet radar [51]. The USSR maintained an aggressive anti-air reconnaissance policy and shot down any aircraft approaching its airspace. In order for US policy makers to obtain imagery intelligence on Soviet military capabilities, especially nuclear capabilities, they would have to fly above Soviet radar. This proved to be a fruitless endeavor, as Soviet radar had advanced beyond the assessments of US military intelligence and possessed the capability to detect and track aircraft above 70,000 feet making the U2 obsolete at the outset.

Focus then shifted from high altitude to stealth by reducing the radar cross-section of the U2. Instead of flying above radar, the plane would avoid radar detection by altering the U2's material composition. Project Rainbow sought to reduce the radar cross section of the U2 by wallpapering a conductive honeycomb grid over the skin of the aircraft to reflect electromagnetic energy [52]. The Rainbow team also investigated altering the surface area characteristics of the U2 through "trapeze" designs to reduce low frequency reflections from the wings. Ultimately the team determined that in order to explore new methods of achieving stealth, the U2 design would need to be scrapped and the entire stealth aircraft concept would need to be reevaluated. Project Rainbow was far from a failure, however, as it provided the foundation for modern stealth technology and inspired the launch into Project OXCART [52].

Pre-Cold War era military strategy was based on raw industrial might and the ability to deliver massive destructive force. The sophisticated air campaigns of WWII and the introduction of rockets, advanced submarines, and nuclear weapons changed that strategic calculus. Modern power projection is not limited by destructive capacity, rather it is limited by ability of the attacker to find enemy targets and avoid countermeasures. The relative advantage of the offense over the defense depends on the ability of the offense to avoid detection. Consequently, the advantage of the defense is incumbent on the ability to detect. This is not to say destructive power is irrelevant, just that it is dependent on unimpeded delivery to the target.

OXCART and associated programs represented a commitment by forward thinking military minds in the US to attaining superiority in the detection arms race resulting in revolution in US stealth and aerial reconnaissance capabilities. One of the most iconic and revolutionary planes in US history, the SR-71 Blackbird, came out of that era of research and development. The Blackbird and its predecessor, the A-12, was the product of engineering innovations in "titanium fabrication, lubricants, jet engines, fuel navigation, flight control, electronic countermeasures, radar stealthiness, and pilot life-support systems." [53]. The Blackbird could fly at speeds of over 2,000 miles per hour and altitudes of over 85,000 feet, both records at the time. The A-12 and SR-71 were integral in intelligence gathering efforts, primarily in assessing enemy missile capabilities. The Blackbird was operational from 1968 to 1989 and played a large role in assessing Soviet military capabilities [54].

At the tail end of the Blackbird's service life, the US introduced its first stealth fighter—the F-1117 Nighthawk. The Nighthawk—also built by Lockheed—built upon and surpassed the stealth capabilities of the Blackbird by introducing 'unstable' design. Aerodynamically unstable aircraft had previously been impossible to fly due to the lack of advanced flight control systems required to keep the plane in the air. Advances in computing technology allowed Lockheed engineers to experiment with radically new 'diamond' pattern designs meant to drastically reduce the radar cross section of the aircraft in an effort to evade increasingly sophisticated Soviet radar and anti-air missile systems [55]. The Nighthawk was designed as an attack aircraft intended to drop guided munitions on high value targets in heavily protected areas. The prime example being Desert Storm, where the Nighthawk was responsible for high value targets like communications stations and other command and control structures [56]. The Nighthawk was pivotal in crippling the Iraqi integrated air defense system in the early stages of the war, one of the most central components to the quick and decisive victory the US achieved. An impressive achievement when the sophistication of the Iraqi IADS system is taken into account—US planners said these high value targets were better defended than Cold War Era Eastern Europe.

The Blackbird and Nighthawk were both stealth aircraft designed to elude enemy detection, however, there is a stark contrast in the effects of these aircraft on the relative stability of the international system. The Blackbird was designed to avoid detection; but its purpose, ostensibly, was to reveal enemy capabilities and movements—an act which reduces the level of US uncertainty and serves to assuage the paranoia of the Cold War security dilemma. Conversely, the Nighthawk increased Cold War tensions because its goal was not to reveal information—rather its goal was to create a strategic liability for the USSR. From this perspective, two stealth technologies can have opposite effects on the stability of the international system. There is, however, another perspective that argues the Nighthawk is a stabilizing deterrent weapon due to its assurance of an American second strike in the event of a nuclear attack. Mutually Assured Destruction, as an agreed upon doctrine, is not viable without the guarantee of a second strike capability. It could also be argued that the Blackbird is destabilizing because the Soviets would never assume the aircraft could or would only be used for reconnaissance purposes, and instead attribute aggressive intentions.

The current generation of stealth bomber is the B2 Spirit—a heavy penetration strategic bomber intended to carry out nuclear strikes on high value targets. However, the changing strategic landscape where the US is engaged in more asymmetrical conflicts necessitated that the B2 be repurposed to deliver a conventional payload. A new generation of Joint Strike fighters—the F22 and F35—also step into the stealth role. Capable of air and ground combat, reconnaissance, and electronic warfare, the joint strike fighters combine the stealth capabilities of the Nighthawk and Blackbird with the power projection capabilities of a modern multipurpose attack aircraft [57].

Aircraft have largely been displaced in their reconnaissance roles by increasingly sophisticated satellite technology. Since in the 1960s, the US has utilized satellite photography to assess enemy capabilities, force movements, and offensive buildups. Early reconnaissance satellites were not capable of matching the level of detail of the SR-71 but eventually the capabilities of satellites obviated the need for a robust aerial reconnaissance program. Satellites are an interesting demonstration of the security dilemma in that they are currently a purely defensive technology, there are no armed satellites. Nevertheless, great powers like the US, China, and Russia have already began preparing for space warfare—satellite mounted missiles, electronic countermeasures, anti-satellite missiles [58]. Like the SR-71, whose purpose was purely informational, states do not have the luxury of taking peaceful intentions for granted.

6 Disrupting the Stealth Paradigm

Are meta-materials a disruptive innovation representing a new paradigm in stealth? At the moment, it is a truly emerging technology. If the pace of R&D in the field continues to progress rapidly, then MMAC could be a game changer in the ensuing years. Below are three scenarios in which MMAC's are judged to have the most potential to disrupt the nature of conflict. An analysis of the effect of meta-materials is shown in Table 1, and the findings are discussed in detail in the following sections.

6.1 Littoral Waters

One of the major markers of modern state versus state warfare is the challenge of Anti-Access Area Denial (A2AD), which is the restriction of movement into (A2) and within (AD) the theatre of conflict. A2AD is not a concept in the history of warfare. A common thread in conflict from the Ancient Greeks to modern America is the desire to deny the adversary at longer and longer ranges. However, A2AD is unique in the short term context because of the fairly unimpeded access enjoyed by the US following the fall of the Soviet Union. From the early 1990's to the late 2000's the US Navy could move into and within virtually any region and "show the flag." Sometimes referred to as "gunboat diplomacy," the tactic is one method for power projection due to the overwhelming advantage the US has over any rival in the maritime environment. In modern times, aircraft carriers give the US global presence and the ability to project power effectively in a crisis situation but potential adversaries have looked to technological innovation to limit the US by imposing unacceptable costs on America projecting power in their regions to enable asymmetric challenges to the US and its allies [59].

A2AD is challenging the paradigm through the use of advanced anti-ship missiles—such as the Chinese (PRC) Deng Feng (DF) 21 and DF-26 intermediate range ballistic missiles [60] and Iranian small boat-mounted cruise missiles and submarine [61–63]. The two threats present different challenges to the US and its allies. The 'carrier killer' is a standoff weapon operating within a complex command, control, intelligence, surveillance, reconnaissance ecosystem (C4ISR) attempting to deny access to US carriers from entering the general vicinity of the South China Sea. The

Table 1 Assessment of stabilizing effects of meta-materials in international security scenarios and over time

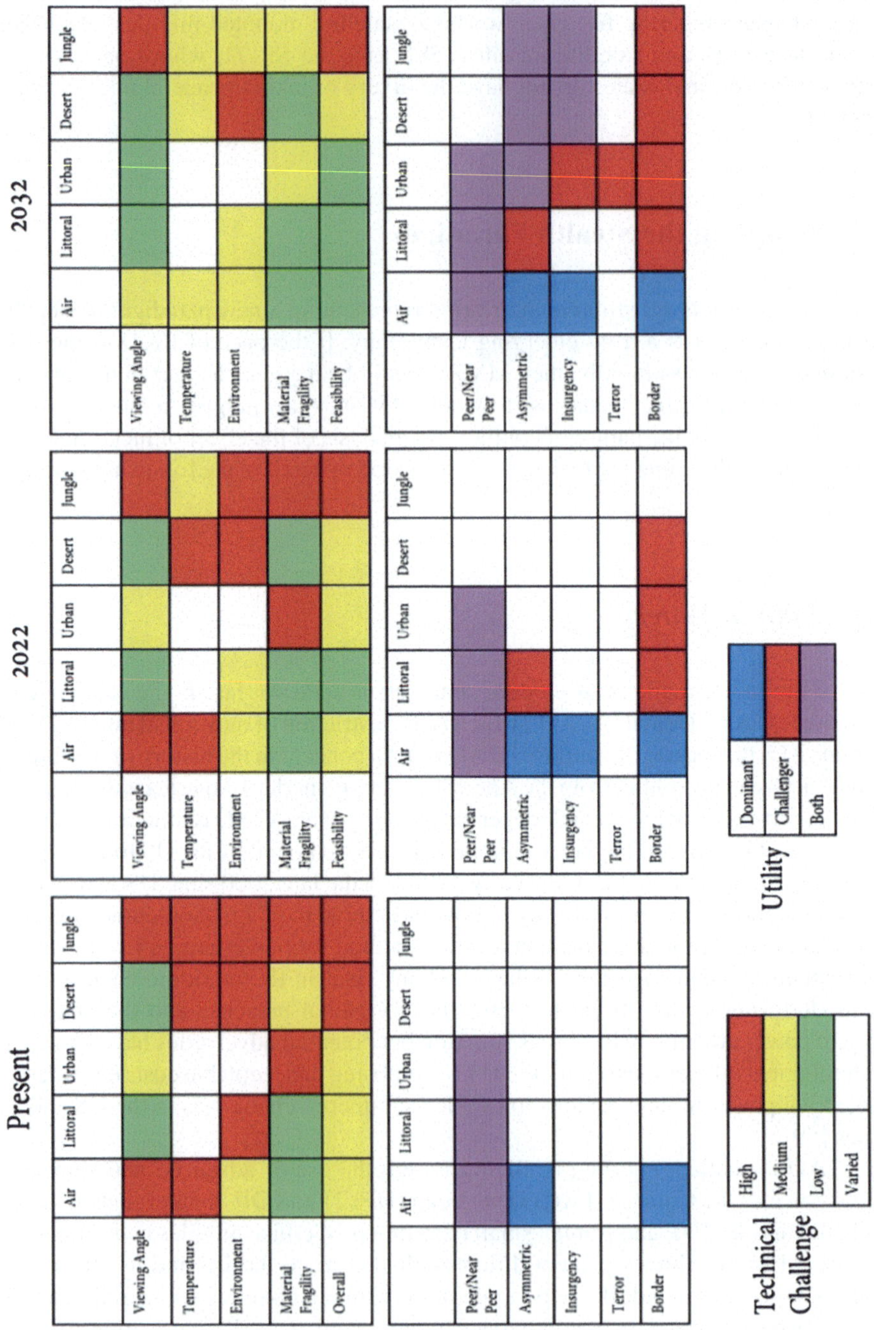

weapons are mounted on trucks or placed on one of the many artificial islands built by China in the past decade in contested waters away from the Chinese coast. US ships routinely patrol through the South China Sea to ensure freedom of navigation and to protect the interests of its Pacific allies, who contest the island building campaign.

The Iranian challenge, however, is on the opposite end of the technological spectrum. The Iranian Navy has equipped what are effectively militarized speedboats—Fast Inshore Attack Craft—with anti-ship cruise missiles [64, 65]. The fast boats are relatively inexpensive and therefore a cost effective means to pursue deterrence through saturation. The conventional fear is that fast boats would swarm and overwhelm an American ship in the Strait of Hormuz—an artery for global energy transportation. While one or a few cruise missiles equipped fast boats would be no match for an American ship, a swarm might be lethal, according to naval wargames [66]. Traditional means for ship to ship combat in this scenario are unfeasible due to cost asymmetry between American Tomahawk cruise missiles and the Iranian fast boats, among other concerns including escalation. The US Navy has investigated the use of lasers and smaller, less expensive missiles as one counter to Iranian fast boats [67].

Metamaterial camouflage could be the complicating factor in either of the two A2AD scenarios. In the hands of the US, the camouflage could potentially ensure freedom of movement into and within the theatre by countering the already precarious Chinese C4ISR capabilities and further complicating Iranian fast boat swarms. Conversely, if the technology were to be utilized by the Chinese or Iranians, the A2AD challenge would be much greater. With the added ability to evade detection of US anti-ship missiles, Chinese vessels would become more brazen in their maneuvers in the South China Sea. Likewise, MMAC would be a force multiplier for Iranian fast boats looking to overwhelm a US ship. The proposed counter measure to fast boats is a ship mounted, laser-guided hellfire missile. Metamaterials could render the laser fire and guidance systems ineffective or severely hampered [68]. In both scenarios, the potential uncertainty created by meta-materials would add to tensions and potentially exacerbate efforts thereby increasing instability.

6.2 Border/Trafficking

A second scenario in which metamaterial camouflage could be deadly is in a border and illicit trafficking situation. The volume of movement of people and goods across borders is higher than at any time in human history due to advances in transportation technology, an interconnected global economy, and ongoing civil conflicts. Illicit movement of people, money, weapons, drugs, and other valuable and stolen goods is also at an all-time high and increasingly complex [69]. Human and narcotics trafficking from the developing to developed world is a public health and human rights crisis. The International Labor Organization estimates that there are over 20 million victims of human trafficking and the industry generates over $150 billion in revenues each year [70]. The global drug trade generates over $450 billion in revenues annually, much of which is fueling civil conflict and organized crime [71].

Opium from the Golden Triangle in South East Asia and the states of Central Europe as well as cocaine from South American Andes is sold to pay for illegal weapons used in conflicts for control of the drug market. Drug conflict causes civil unrest and sows the seeds for civil conflict. In Afghanistan, opium stocks the coffers of corrupt politicians and government officials as well as war lords and terrorist organizations like the Taliban. These organizations not only impede local development but also export terrorism to the developed world.

Although initially metamaterial technology will be available to only the militaries of the most sophisticated countries, it likely will diffuse to smaller states and non-state actors decades in the future. If a criminal trafficking organization could obtain an invisibility capability, their ability to covertly cross borders with illicit goods would be greatly enhanced. According to the Department of Homeland Security, the US already struggles to interdict maritime and ground based trafficking efforts [72]. Additionally, if the technology finds useful and profitable commercial applications, diversion or co-option of technology for malevolent purposes is an increasing concern. The addition of invisibility further complicates the efforts of border authorities to successfully detect and interdict illicit crossings.

Perhaps a more worrying scenario than drugs or people is the trafficking of weapons of mass destruction across borders. A chemical, biological, or nuclear weapon crossing a border or into a protected area of a city in order to attack a high value target would be a dream for a terrorist organization. Metamaterials are not likely to disrupt the majority or the most important WMD detection technologies, which do not rely on vision for detection. There are multiple technical and operational scenarios to be addressed that may create vulnerabilities, however.

Regardless the entire border and diplomatic security paradigm is underpinned by the ability to visually perceive the threat space. In this scenario, the US has no advantage in having metamaterials but is faced with a significant threat if the technology were to fall into the hands of an actor with both the means to operationalize the technology and motive to use it against high value targets or for trafficking illicit goods. The asymmetric advantage falls largely to an adversary in such a scenario.

7 Conclusions

There are two lessons learned from the history of stealth technology. First, no technology or capability exists in a vacuum; an equally monumental advance in steal or counter stealth could have totally different repercussions based on the state of international politics and the weapons systems the technology supports. Stealth is not an objective in itself, stealth is a means to an end, a support technology used to either gather or conceal information from the enemy in order to gain an advantage in conflict. Early stealth aircraft were stabilizing because they revealed information about the non-aggressive intentions of the USSR. However, the same aircraft were destabilizing later on because of the liability it presented to air defense systems. Second, if one state is developing a capability then its peers are developing a counter capability. Innovations in stealth aircraft, satellites, and submarines are mirrored by the development of radar, missiles, mines, and jamming technology. States will go to great expense to avoid an imbalance in capabilities and preemptively develop counter measures to mitigate the risk being overpowered.

What does this mean for metamaterial adaptive or active camouflage? It would be unwise to assume metamaterial cloaking will be developed unchecked by the concurrent development anti-cloaking countermeasures, as evidenced by the historical development of stealth and counter-stealth technology. Even the state leading the development of metamaterials—the United States—cannot risk a critical vulnerability to their own capability because it will not always be their own exclusively. The development of metamaterial cloaking also depends on if the status quo powers even desire such a capability.

One immediate security implication is the importance of policy efforts to control access to the technology as much as possible. In the long term this is not a viable strategy due to confounding nature of a globally-connected economy, characterized by digitally-enabled exchange of information among coordinated de-centralized entities, and scientific progress. Those challenges are hardly limited to metamaterials: export controls are already challenged by emerged and emerging technologies. This capability is likely to further exacerbate those challenges and ways that have been observed previously and possibly in newly disruptive ways. As an interim approach, efforts in 'tech monitoring' and awareness can help bridge the gap between the time when metamaterial camouflage is developed and when it becomes a strategic liability. In the long run, since the technology does favor challengers of the US more than the US as a status quo power, it would be beneficial to emphasize countermeasure development concurrently with the development of the technology itself.

The strengths of status quo powers lie in their ability to project power with large weapons systems like missiles, aircraft, armor, and naval vessels. Advances in stealth at the personal and small vehicle level, the areas which metamaterials are potentially most disruptive, asymmetrically benefit non-state actors seeking to disrupt national security through terrorism, trafficking, insurgencies, and hybrid (or gray zone) warfare. Of course, there are applications where invisibility would be beneficial for status quo powers—such as covert action and special operations—but the liabilities created by individual invisibility may outweighs the benefits. Given the nature of the international system, some state or non-state actor will eventually have both the incentive and capability to develop metamaterial cloaking in order to increase their own power. Therefore, intentionally staunching its development, such as through preemptive restrictive treaties, would also be unwise. To mitigate the risk of proliferated metamaterial cloaking, status quo powers should seek to develop counter-cloaking technology at a faster pace than cloaking technology.

References

1. McMaster HR Jr (2016) Maintaining overmatch: the future of combat vehicle modernization. Association of the United States (AUSA) Institute of Land Warfare Hot Topic: Army Ground Combat Platforms Army, 1 Nov 2016. https://www.ausa.org/events/hot-topic-army-platforms/sessions/keynote
2. Kosal ME (2016) Future operational environment and movement and maneuver: disruptive technology for mobility, protection, and firepower. Association of the United States (AUSA) Institute of Land Warfare Hot Topic: Army Ground Combat Platforms Army, 1 Nov 2016, Arlington, VA. https://www.ausa.org/events/hot-topic-army-platforms/sessions/panel-discussion-1
3. Kosal ME (2018) Emerging chemical and biological technologies: security & policy challenges. In: Tetris C et al (eds) Responsible conduct in chemistry research and practice: global perspectives. Oxford University Press, pp 51–68
4. Zheludev NI (2010) The road ahead for metamaterials. Science 328(5978):582–583
5. Shang L, Zhang W, Ke X, Zhao Y (2019) Bio-inspired intelligent structural color materials. Mater Horiz 6(5):945–958
6. Kosal ME (2009) Nanotechnology for chemical and biological defense. Springer Academic Publishers, New York
7. Alitalo P, Tretyakov S (2009) Electromagnetic cloaking with metamaterials. Mater Today 12:22–29
8. Chen H, Chan CT, Sheng P (2010) Transformation optics and metamaterials. Nat Mater 9:387–396
9. Han T et al (2014) Full control and manipulation of heat signatures: cloaking, camouflage and thermal metamaterials. Adv Mater 26:1721–1734
10. Bilotti F, Tretyakov S (2013) Amorphous metamaterials and potential nanophotonics applications. In: Rockstuhl C, Scharf T (eds) Amorphous nanophotonics. Springer, Berlin, Heidelberg, pp 39–66
11. Di Falco A, Ploschner M, Krauss TF (2010) Flexible metamaterials at visible wavelengths. New J Phys 12:11306–11313
12. Rogers JA, Someya T, Huang Y (2010) Materials and mechanics for stretchable electronics. Science 327:1603–1607
13. Armstrong S (2012) Biophotonics: photonic crystals aid fish's night vision. Nat Photonics 6:575
14. Wang N, Zhao Y, Jiang L (2010) Bioinspired synthesis and preparation of multilevel micro/nanostructured materials. Front Chem China 5:247–261
15. Stuart-Fox D, Moussalli A (2009) Camouflage, communication and thermoregulation: lessons from colour changing organisms. Philos Trans R Soc B 364(1516):463–470
16. Greenleaf A, Kurylev Y, Lassas M, Uhlmann G (2009) Cloaking devices, electromagnetic wormholes, and transformation optics. SIAM Rev 51(1):3–33
17. Aranovich VM et al (2004) Linear and nonlinear wave propagation in negative refraction metamaterials. Phys Rev B 69:165112
18. Veselago VG (1968) The electrodynamics of substances with simultaneously negative values of ε and μ. Sov Phys Usp 10:509–514. https://ufn.ru/ru/articles/1967/7/d/
19. Pendry JB, Schurig D, Smith DR (2006) Controlling electromagnetic fields. Science 312(5781):1780–1782
20. Gabrielli LH, Cardenas J, Poitras CB, Lipson M (2009) Silicon nanostructure cloak operating at optical frequencies. Nat Photonics 3(8):461–463
21. Smolyaninov II, Smolyaninova VN, Kildishev AV, Shalaev VM (2009) Anisotropic metamaterials emulated by tapered waveguides: application to optical cloaking. Phys Rev Lett 102:213901
22. Lee JH, Blair J, Tamma VA, Wu Q, Rhee SJ, Summers CJ, Park W (2009) Direct visualization of optical frequency invisibility cloak based on silicon nanorod array. Opt Express 17(15):12922–12928

23. Valentine J, Li J, Zentgraf T, Bartal G, Zhang X (2009) An optical cloak made of dielectrics. Nat Mater 8(7):568–571
24. Ergin T, Stenger N, Brenner P, Pendry JB, Wegener M (2010) Three-dimensional invisibility cloak at optical wavelengths. Science 328(5976):337–339
25. Ni X, Wong ZJ, Mrejen M, Wang Y, Zhang X (2015) An ultrathin invisibility skin cloak for visible light. Science 349(6254):1310–1314
26. Xu Y et al (2018) Light–matter interaction within extreme dimensions: from nanomanufacturing to applications. Adv Opt Mater 6(18):1800444
27. Xu C, Stiubianu GT, Gorodetsky AA (2018) Adaptive infrared-reflecting systems inspired by cephalopod. Science 359(6383):1495–1500
28. ADAPTIV—Cloak of Invisibility (2019) BAE Systems PLC. http://www.baesystems.com/en-us/feature/adativ-cloak-of-invisibility
29. Merritt R (2013) Do-it-yourself invisibility with 3-D printing. Center for Metamaterials and Integrated Plasmonics Newsletter, 6 May 2013. http://metamaterials.duke.edu/news/do-it-yourself-invisibility-3-d-printing
30. Technology 2019. Metamaterials Commercialization Center. http://www.metamaterialscenter.com/applications/imaging/
31. Schittny R, Kadic M, Bückmann T, Wegener M (2014) Invisibility cloaking in a diffusive light scattering medium. Science 345(6195):427–429
32. Smith DR (2014) A cloaking coating for murky media. Science 345(6195):384–385
33. Nye JS (1990) Soft power. Foreign Policy 80:153–171
34. Tang Shiping (2009) The security dilemma: a conceptual analysis. Secur Stud 18(3):587–623
35. Mearsheimer JJ (2001) The tragedy of great power politics. Norton, New York
36. Jervis R (1978) Cooperation under the security dilemma. World Polit 30(2):167–214
37. von Clausewitz C (1832) On war
38. Goldstein A (2000) Deterrence and security in the 21st century. Stanford University Press, Stanford, Calif
39. Treaty Between the United States of America and the Union of Soviet Socialist Republics on the Limitation of Anti-Ballistic Missile Systems (1972) US Department of State, Other Releases, Bureau of Arms Control, 26 May 1972. https://www.state.gov/t/isn/trty/16332.htm
40. Defense Science Board (2007) 21st century strategic technology vectors, vol I, main report. Office of the Under Secretary of Defense for Acquisition, Technology, and Logistics, Washington, D.C., Feb 2007. https://www.acq.osd.mil/dsb/reports/2000s/ADA463361.pdf
41. Maxwell JC (1865) A dynamical theory of the electromagnetic field. Phil Trans R Soc Lond 155:459–512
42. Hertz H (1887) On electromagnetic effects produced by electrical disturbances in insulators
43. Bauer AO (no date) Christian Hülsmeyer and about the early days of radar inventions. Foundation Centre for German Communications and Related Technologies. http://aobauer.home.xs4all.nl/Huelspart1def.pdf
44. Page RM (1962) The early history of radar. Proc IRE 50(5):1232–1236
45. Skolnik MI (1985) Fifty years of radar. Proc IEEE 73(2):182–197
46. Allison DK (1981) New eye for the navy: the origin of radar at the Naval Research Laboratory. NRL report S466, Sept 1981. https://apps.dtic.mil/dtic/tr/fulltext/u2/a110586.pdf (See also "Development of the Radar Principle" U.S. Naval Research Laboratory. https://www.nrl.navy.mil/accomplishments/systems/radar/)
47. Latham C, Stobbs A (2006) The birth of British radar: the memoirs of Arnold 'Skip' Wilkins. Speedwell, Birmingham
48. Baldwin S (1932) "International Affairs" Speech in house of commons, UK Parliament, 10 Nov 1932, vol 270, pp 525–564. https://api.parliament.uk/historic-hansard/commons/1932/nov/10/international-affairs
49. Clark GC (1997) Deflating British radar myths of world war II. Air Command and Staff College. https://fas.org/man/dod-101/ops/docs/97-0609F.pdf
50. Stone M, Banner GP (2000) Radars for the detection and tracking of ballistic missiles, satellites, and planets. Linc Lab J 12(2). https://pdfs.semanticscholar.org/477d/7311053bd57e12d74355548c83e770420ad6.pdf

51. Pedlow GW, Welzenbach DE (1998) The CIA and the U-2 program, 1954–1974. History Staff, Center for the Study of Intelligence, Central Intelligence Agency, 1998. https://www.cia.gov/library/center-for-the-study-of-intelligence/csi-publications/books-and-monographs/the-cia-and-the-u-2-program-1954-1974/u2.pdf
52. The U-2's Intended Successor: Project OXCART, 1956–1968. The National Security Archive. http://nsarchive.gwu.edu/NSAEBB/NSAEBB434/docs/U2%20-%20Chapter%206.pdf
53. OXCART vs Blackbird: do you know the difference? Central Intelligence Agency. https://www.cia.gov/news-information/featured-story-archive/2015-featured-story-archive/oxcart-vs-blackbird.html. Accessed 18 Nov 2015
54. Roblin S (2016) The SR-71 blackbird: the super spy plane that outran missiles. Natl Interes. http://nationalinterest.org/blog/the-sr-71-blackbird-the-super-spy-plane-outran-missiles-18142
55. Majumdar D (2015) The ultimate comeback: should the lethal F-117 stealth fighter fly again? Natl Interes. http://nationalinterest.org/blog/the-buzz/the-ultimate-comeback-should-the-lethal-f-117-stealth-14180
56. Operation Desert Storm: Evaluation of the Air Campaign. NSIAD-97-134, United States Government Accountability Office, 12 June 1997. https://www.gao.gov/products/NSIAD-97-134
57. The F-35 Lightning II (2017) The Joint Strike Fighter Program. http://www.jsf.mil/
58. Billings L (2015) War in space may be closer than ever. Sci Am. https://www.scientificamerican.com/article/war-in-space-may-be-closer-than-ever/
59. Using clever technology to keep enemies at bay. The Economist (UK). https://www.economist.com/special-report/2018/01/25/using-clever-technology-to-keep-enemies-at-bay
60. Jeong-ho L (2019) China releases footage of 'Guam killer' DF-26 ballistic missile in 'clear message to the US. South China Morning Press. https://www.scmp.com/news/china/military/article/2183972/china-releases-footage-guam-killer-df-26-ballistic-missile-clear
61. Krepinevich AF, Watts B (2003) Meeting the anti-access and area-denial challenge. Center for Strategic and Budgetary Assessments (CSBA), Washington, D.C., 20 May 2003. https://csbaonline.org/research/publications/a2ad-anti-access-area-denial
62. Iranian naval forces: a tale of two navies. Office of Naval Intelligence, Feb 2017. https://www.oni.navy.mil/Portals/12/Intel%20agencies/iran/Iran%2002221 7SP.pdf
63. Egozi A (2019) Iranian missile misfires as other attacks hit home. Break Def. https://breakingdefense.com/2019/06/iranian-missile-misfires-as-other-attacks-hit-home/
64. Iran's naval forces: from Guerilla warfare to modern naval strategy. Report from the Office of Naval Intelligence, Washington DC, Feb 2009. http://www.fas.org/irp/agency/oni/iran-navy.pdf
65. Singh A (2010) Dark chill in the Persian Gulf—Iran's conventional and unconventional naval forces. Marit Aff J Natl Marit Found India 6:108–132
66. Millennium Challenge 02 (2002) United States Joint Forces Command. http://web.archive.org/web/20070928005405/. http://www.jfcom.mil/about/experiments/mc02.htm
67. Keck Z (2015) America's master plan to crush Iranian warships. Natl Interes. http://nationalinterest.org/blog/the-buzz/americas-master-plan-crush-iranian-warships-13602
68. Young A, Barker D, Owens W (2012) Methods and systems for optical focusing using negative index metamaterial. US Patent 8180213 B2, May 2012
69. Finklea KM (2013) Organized crime: an evolving challenge for U.S. law enforcement. Congressional Research Service, Report No. R41547. https://www.everycrsreport.com/reports/R41547.html
70. International Labor Organization (2012) New ILO global estimate of forced labour: 20.9 million victims. http://www.ilo.org/global/about-the-ilo/newsroom/news/WCMS_182109/lang--en/index.htm
71. Drug Trafficking (2019) United Nations Office on Drugs and Crime. https://www.unodc.org/unodc/en/drug-trafficking/
72. Nixon R (2017) Coast guard faces challenges at sea, and at the budget office. New York Times. https://www.nytimes.com/2017/07/04/us/politics/coast-guard-faces-challenges-at-sea-and-at-the-budget-office.html

Attempting to Predict the Proliferation of Lethal Autonomous Weapons Systems: A Statistical Analysis

Sara M. Schmitt

Abstract Lethal Autonomous Weapons Systems (LAWS), while much-discussed in research throughout the academic and political realms, are often highly politicized, with people of every side of the argument claiming vastly different things about the capability, legality, and morality of these systems. This research seeks to dispel some of the uncertainty surrounding these weapons systems by compiling evidence from academics, philosophers, and military officials to create a holistic briefing. This accomplished, the research draws upon Stephen Meyer's *The Dynamics of Nuclear Proliferation* to predict the proliferation of lethal autonomy by examining shared characteristics among the six states currently confirmed to be in possession of this technology. The research results in an algorithm with 64.5% explanatory power over the variation in possession of lethal autonomy, which can be used to predict with reasonable accuracy a given state's likelihood to proliferate. The algorithm could serve organizations seeking to track proliferation by functioning as a mathematical flashlight to highlight key warning signs of imminent acquisition, thereby protecting against destabilizing shifts in the balance of power status quo.

1 Introduction

Since the invention of the computer in the early 20th century, the possibility of machine intelligence has captivated the minds of science and the arts alike. Movies like *Wall-E*, *RoboCop*, and *Transformers* have explored the possibilities of autonomous intelligence and the saving or destroying powers of technology made in the image of humankind. While these movies are entertaining, their plausibility is often dismissed or equated to the scientific possibility of aliens—not entirely impossible, but still beyond the reach of modern-day science. This dismissal is incorrect, as such autonomous technology is within the realm of scientific possibility and is present in the arsenals of several global superpowers. While these inventions are not

S. M. Schmitt (✉)
Georgia Institute of Technology, Atlanta, GA, USA
e-mail: sschmitt8@gatech.edu

M. E. Kosal (ed.), *Disruptive and Game Changing Technologies in Modern Warfare*, Advanced Sciences and Technologies for Security Applications,
https://doi.org/10.1007/978-3-030-28342-1_9

as humanoid as the robots of *Transformers*, Lethal Autonomous Weapons Systems (LAWS) are shrouded in as much sensationalism and uncertainty as their fictitious counterparts.

For the purpose of this research, LAWS will be defined as those weapons systems capable of executing an assigned task without human contact or intervention after the initial launch. The security implications of LAWS technology are numerous; a successful system could carry out an attack anywhere in the world with a guarantee of zero fatalities for the operating entity. Additionally, the ability to perpetrate warfare with zero military casualties is an immense advantage over actors affected by limited troop numbers and public opinion over fatalities. With this kind of an advantage at stake, the proliferation of these weapons is a topic worth tracking and, if possible, predicting to understand how proliferation might affect the balance of power status quo. This research seeks to do so by exploring a series of related questions. What are the key factors in determining whether or not a state will develop lethal autonomous weapons systems? Can these factors be used to predict the proliferation of this technology? If so, which states are most likely to acquire this technology in the future? Through a process that will be described subsequently in detail, the author finds that the possession of nuclear weapons, research and development spending, and military spending are the best indicators of LAWS possession, but the formula created from these variables is insufficiently accurate to have a predictive capacity.

2 Relevance and Importance

An algorithm predicting the proliferation of autonomous weapons systems provides a strategic advantage for the United States that reflects the strategic advantage of LAWS possession. The advantage of possessing LAWS while fighting against a state (or non-state actor) that lacks this technology is significant; the ability to carry out a substantial and sustained threat without risking the lives of troops provides a significant battlefield advantage and allows any single campaign to be prolonged further than public opinion would normally allow given the publicity surrounding troop deployment (specifically for the United States and many of its allies).

Furthermore, LAWS technology has a meaningful domestic utility as well. State necessities like border patrol and many aspects of law enforcement could be executed for a longer period of time and in more hazardous conditions with autonomous vehicles than without and with arguably the same attention to detail. In fact, combatants are known to have surrendered to armed unmanned autonomous vehicles (UAVs or drones) in the past. This phenomenon was first recorded in the 1991 Gulf War, when several Iraqis waved a white flag at a damage-assessment drone to signal their surrender [1, p. 279]. It is not unreasonable to assume that such a phenomenon could also occur in domestic law enforcement with an autonomous robot.

From a threat assessment perspective, LAWS could also be programmed for assassinations, the suppression of human rights, or targeted killing based on ethnicity. For this reason, these powerful and versatile weapons systems may be a threat to international strategic stability with some comparisons to nuclear technology—capable of great domestic good (sustainable energy, in the case of nuclear technology) or great international destruction (as in the case of a hydrogen bomb). Keeping a close eye on the potential proliferation of this technology should be an important priority.

The goal of this research—creation of an algorithm capable of predicting the next most-likely possessor of LAWS technology—could be a useful solution to questions surrounding the proliferation of autonomous weapons. This type of algorithm has the potential to be somewhat of a balancing factor in the global drive to modernize warfare and advance existing defense-sector technologies, as it could potentially predict future proliferation and foresee changes in the status-quo. It could also serve organizations seeking to track the proliferation of this new technology by functioning as a mathematical flashlight to highlight key warning signs of imminent acquisition. As this technology manifests itself throughout the global defense arena, the implications of such an algorithm will become more apparent; today, however, noting the potential utility of this algorithm is sufficient.

3 Background

Despite the fact that lethal autonomous weapons systems are relatively new compared to many other military technologies, there is a wealth of information that is relevant to this research. This chapter presents background on LAWS with an emphasis on how it relates to the research conducted. First the history of LAWS technology and a description of the state of LAWS possession today is reviewed. Next, experts' opinions on the technology, which provide two alternative uses for monitoring and predicting the proliferation of LAWS, are explored.

3.1 About Lethal Autonomous Weapons Systems

Removing the human element from warfare is not a new concept, as weapons designers have consistently sought to provide advantage to the user by keeping people as far from the point of conflict as possible. Firearms and artillery provided an advantage over swords by supplementing hand-to-hand combat, while armed aircraft provided a similar advantage by placing the operator out of reach of ground combatants. Drones were another step in this direction, as suddenly pilots were removed from the cockpits of airplanes and a system could monitor a dangerous situation or fire lethal ballistics without direct physical threat to its operator. The advantageous nature of this technology is clear; under the Obama administration, drone strikes increased by a factor of ten as compared with the Bush administration—and this is despite President Obama's campaign rhetoric to limit the U.S. presence in the Middle East [2].

In a notional sense, lethal autonomous weapons systems take this process one step further by removing the pilot from the decision-making process. Instead of an airman or other uniformed service member sitting at a control board and operating the drone, a computer scientist writes code for a corresponding robot—teaching it to make decisions based on certain stimuli—and, it is released to carry out its assigned task by its initial operator. Hypothetically, no further contact may occur until the robot returns and is re-programmed or released in a different area.

As of 2017, the Campaign to Stop Killer Robots declares six states to be in the process of developing lethal autonomous weapons systems based on corporate and governmental press releases and demonstrated weapons capabilities. These states are China, Israel, Russia, South Korea, the United Kingdom, and the United States [3]. Although widespread LAWS technology is a relatively new phenomenon within the history of weapons manufacture, it is prevalent in several militaries worldwide. Examples of this technology include the Phalanx Close-in Weapons System (U.S. Navy); the Tomahawk Cruise Missile (U.S. Navy); the ARV (Armed Robotic Vehicle; developed by Lockheed Martin, but cancelled in 2011 due to mobility concerns); Predator Unmanned Aerial Vehicles (UAVs) equipped with hellfire missiles; the TALON SWORDS platform (developed by Foster-Miller/QinitiQ); the Modular Advanced Armed Robotic System (MAARS); the Phalanx system for Aegis-class cruisers (U.S. Navy); the MK-60 encapsulated torpedo (CAPTOR) sea mine system (U.S. Navy); and the Patriot anti-aircraft missile batteries [1, pp. 276–278].

These systems represent the forerunners of lethal autonomy given their ability to detect the conditions of their environment and react using lethal force. For instance, the Phalanx system is described by the U.S. Navy as "the only deployed close-in weapon system capable of autonomously performing its own search, detect, evaluation, track, engage and kill assessment functions." [4]. The TALON SWORDS platform lists "autonomy and mapping" as a system skill, but also incorporates multiple options for direct remote command, indicating the ability to function autonomously or under direct command [5]. Anti-tank landmines also present an early example of lethal autonomy given their judgment of a situation (for example, whether a change in pressure exerted from above is associated with a vehicle or a smaller source) and respond accordingly (for example, explode or remain dormant). It is important to note that with each of these systems, a human remains in command of deployment of lethal force but does not play a direct role in supervising the system.

Although the systems discussed above are exclusively representative of U.S.-developed technology, other states are currently in the process of developing their own manifestations of lethal autonomy. The Taranis UAV—a U.K. developmental project lead by BAE Systems—was announced in 2006, at which time BAE Systems claimed that the weapons system would build upon years of research into "full autonomy." [6]. Despite this early announcement, BAE Systems describes Taranis as designed to operate "under the control of a human operator," indicating differing levels of autonomous capabilities [7]. Russia has been less opaque with its intention to develop fully autonomous weapons systems. In July 2017, Kalashnikov's Director for Communications announced to TASS news agency the completion of a "fully automated combat module based on neural network technologies that enable it to identify targets and make decisions." [8].

3.2 Arguments Regarding Lethal Autonomy

As with many new technologies, lethal autonomous weapons have generated controversy in both the scholarly and policy worlds; however, the conversation surrounding LAWS has been particularly heated and has drawn commentators from other sectors as well. This discussion is important because it helps to place the utility of a prediction algorithm into context. In this section, three broad views are presented and their relationship to a potential proliferation-prediction algorithm is discussed. Although each position is multi-faceted and widely supported, for the sake of brevity, this document will highlight only the most critical supporting arguments from each perspective.

The first perspective on lethal autonomy is one of optimism and confidence in the technological capabilities of the system. This opinion is well-summarized in chapter thirteen of *New Technologies and the Law of Armed Conflict*, in which author Jeffrey S. Thurnher articulates several key points regarding the legality of lethal autonomous weapons, stating that autonomous weapons systems "may even be able to provide greater protections to civilians by delivering more precise and accurate strikes." [9, p. 215]. Proponents of this argument assert that an autonomous system can be programmed to learn from its experiences, much as humans do [10]. Furthermore, they argue, LAWS could make better, more objective decisions than soldiers due to an impartiality not compromised by adrenaline or fear [11]. These systems also have a technological advantage over remotely piloted vehicles (also known as 'drones'), as there are no electronic control signals for an adversary to block or within which poor conditions or geography could cause interference.

The ability of LAWS technology to comply with the standards has been described in international law [9, pp. 219–226]. Accordingly, two aspects of the international legal code apply to this new technology—weapons law and targeting law. Thurnher finds that LAWS are hypothetically discriminate weapons systems and can be designed to avoid superfluous injury, making them compliant with international law (although he does note that each state should perform a legal review prior to deployment of any new weapons system). The second legal aspect—targeting law—is broken down into three parts: distinction, proportionality, and precautions. Thurnher declares that in this case, too, LAWS can be programmed with enough precision as to make them compliant with the standards described in these laws.

The second perspective, however, takes a different view of asserting that technology cannot provide the needed fidelity and precision for lethal judgements thereby questioning the legality of such weapons systems. Leading scholar, Noel Sharkey, argues in "The Evitability of Autonomous Robot Warfare" that programming a machine to respond proportionally is incredibly difficult given the number of possible outcomes of any given situation [12]. Furthermore, autonomy opponents argue that in the case of a violation of international humanitarian law by an autonomous weapon, culpability is a major issue. If an autonomous weapon commits a war crime, who can be held responsible? Additionally, the lack of human emotions that make autonomous weapons reliable may also makes them dangerous, as they are incapable of acting upon a moral code. Essentially, this perspective posits that lethal autonomy is too unreliable to trust with life-and-death decisions in combat.

The third perspective stems from a more practical standpoint, as military experts approach the issue of lethal autonomy from a non-civilian mindset and seek to assess its utility within the realm of national security. According to this perspective, as summarized in the 2016 Joint Concept for Robotic and Autonomous Systems, the proliferation of lethal autonomy is not an issue of legality or morality, but one of time and technological advancement [13, p. 4]. This vision of future robotic and autonomous systems (RAS), as articulated succinctly by Wayne P. Hughes, US Navy (ret) of the Nabal Postgraduate school, is that "[t]he fundamental error in a debate over robotic development is to think that we have a choice. This world is coming, rapidly coming." [14]. Based on this assumed necessity and inevitability, the U.S. military vision for RAS posits a combination of human and RAS deployment in which RAS provide logistical support, or augment and execute particularly high-risk assignments in order to lessen risk to humans.

Although this perspective views proliferation of LAWS as inevitable, it does address specific concerns raised in previous arguments for or against LAWS. The military experts who contributed to this document list many of the same arguments for the use of RAS systems as civilian experts, including higher performance standards, increased situational awareness, and faster response times. The document clearly states that all systems must abide by the Law of Armed Conflict, though it also cites a need for adaptive laws and norms as technological capabilities outpace the language of existing frameworks [13, p. 7].

These three arguments are important because they imply different interpretations of utility for the predictive algorithm. In the case of the first perspective, Thurnher could find utility in the algorithm in that it would allow military balancing and maintenance of the status-quo balance of power. Alternatively, Sharkey could use the algorithm to determine which countries to target in the campaign to ban LAWS technology. For military experts, the algorithm is useful in that it allows foreknowledge of what is seen as inevitable proliferation. Thus, the algorithm is useful to a broad array of professionals on every side of the issue. This versatility shows a lack of bias in the algorithm, since it is applicable to all sides of an otherwise highly divisive argument.

4 Methodological Approach

This research seeks to answer one main question and two dependent sub-questions: what are the key factors in determining whether or not a state will develop lethal autonomous weapons systems (LAWS)? Can these factors be used to predict the proliferation of this technology? If so, which states are most likely to obtain LAWS in the future? The operating hypothesis of the research states that if a state possesses LAWS technology, the state has high research and development spending due to the technical complexity of fully-autonomous artificial intelligence. Other important factors considered include high military spending and alliances with other states which possess LAWS, since military cooperation, joint drills, and commercial ties could potentially increase the rate of technological proliferation among allies.

Among the goals in conducting this research is to take a controversial topic and increase the understanding of its implications for national and international security. The framework utilized here, although primarily designed to be predictive in nature, can also be explanatory in divulging why a given state chooses to pursue autonomous weaponry.

Looking to the literature and previous frameworks for explaining the proliferation of high technologically-based weapons, i.e., nuclear weapons, this research employed a previously published method as the basis for the model's analysis. The method is adapted from Stephen Meyer's *Dynamics of Nuclear Proliferation* and serves as the foundational inspiration for the research [15].

Meyer's work *The Dynamics of Nuclear Proliferation* seeks to explain the proliferation of nuclear weapons by examining two defining aspects of the technology and its spread throughout the world. First, he looked at the technological imperative of nuclear weapons, which he defines as the compulsion to "go nuclear" or produce a nuclear weapon based upon the ability to do so [15]. Meyer then tests the motivational hypothesis by examining several different factors that could encourage a state to obtain nuclear weapons.

Beginning his investigation into the reasoning behind the proliferation of nuclear weapons, Meyer examines at the technological imperative of nuclear development. In this portion of his research, Meyer rephrases the chicken and egg dilemma by asking which comes first—the desire to possess a nuclear weapon or the technical ability to realize such an ambition. In a series of case studies testing different models of proliferation, Meyer looks at countries such as Japan who have the technological ability to develop a nuclear weapon yet, for various reasons, choose not to develop these weapons. As the work states, he is "unable to find consistent support for any of the models of the technological imperative," or technological determinism [15, p. 88]. Meyer is able to reject the hypothesis of the technological imperative, stating that the ability of a state to produce a given weapon does not inherently require the state to do so. On the basis of these findings, this research also rejects the technological imperative as a factor in proliferation and excludes production capability as a determining factor in the analysis.

Next the political and security-driven motivation hypothesis of nuclear proliferation via multivariate statistical analysis was tested. The list of variables Meyer used in his evaluation are global power status/pretensions; pariah status; military alliance with a nuclear power; security threat from a nuclear power; adversary with latent capacity; overwhelming conventional military threat; domestic turmoil; major military defeat and national self-image; regional proliferation; intolerable economic defense burden; and other factors not included in these categories. In addition to these political factors, Meyer also examines variables that might dissuade or disincentivize a state to pursue nuclear weapons, such as presence of an alliance with a nuclear power, international legal commitments, a state rival with a latent capacity, risk of unauthorized seizure, preemptive intervention by a major power, and a desire to maintain a peaceful reputation. As will be discussed later, several of these factors (both motivational and dissuasive) are considered in this research or are adapted to align more closely with potential proliferation of lethal autonomous weapons systems.

The variables listed above are tested via a series of statistical analyses to determine their individual relevance to nuclear proliferation. After analyzing these variables in this manner, he seeks to apply them to states to predict the risk of any given state choosing to produce or otherwise obtain nuclear weapons. He denotes each analyzed state as a hard-, middle-, or soft-core proliferant based upon likelihood to proliferate, thereby giving his calculations a real-world application [15, pp. 160–162]. This research replicates Meyer's process, using many of the same variables to attempt to predict proliferation of lethal autonomous weapons systems.

Stephen Meyer's research is relevant to this work in a general sense in that both works attempt to explain the proliferation of a specific technology by examining factors common to the states already in possession at the time of the research. Furthermore, the two works use quantitative analysis to assess an international security question and rely heavily on statistical analysis as a tool and gauge for robustness of the results.

The main deviation between these two works stems from the end result. Where Meyer tests individual variables for relevance to nuclear weapons proliferation, this research tests individual variables and then seeks to combine the most relevant into a comprehensive algorithm that may assist policymakers and others in forecasting potential proliferation of autonomous weapons systems.

Although Meyer's design is an excellent starting point for a model to assess proliferation of lethal autonomy, it requires a few changes to reach optimum utility. Whereas Meyer's sample for analysis encompassed all states then-currently possessing nuclear weapons, the present analysis used a larger sample size to determine which factors are not critical indicators of possession of LAWS technology and then rejected these factors from the resulting model. The sample for this research is comprised of the states possessing the 20 most powerful militaries in the world based on the rankings from the Global Fire Power Military Strength Ranking, which utilizes over 50 factors to determine each state's power index score [16]. This research used the 2017 rankings to identify the 20 states with the most powerful military rankings for analysis. The sample also includes all six states identified as possessing states by the Campaign to Stop Killer Robots.

After determining the sample to be used for analysis, the author then conducted research to determine which specific variables to include as independent variable leading to possession of lethal autonomous weapons systems. In total, the author began with nine independent variables in order to statistically analyze their effects on the dependent variable of LAWS possession. Figure 1 shows each variable chosen for the analysis, which will be discussed further in a subsequent section.

The qualitative variables associated with each country are numericized for the statistical analysis. Thus, the dependent variable is assigned one of two numbers—a one indicates possession of LAWS and a zero indicates lack of possession of LAWS. LAWS possession is determined by a list provided by the Campaign to Stop

Rank	State	LAWS	Allied with LAWS state	Military Spending	R&D Spending	Science Test Rank	Nuclear Weapons	CCW Signatory	Armed Forces Personnel	Robot Density	Bureaucracy
1	U.S.A	1	Y - NATO	611.186	2.79	25	1	0	2,363,675	176	75.7
2	Russia	1	Y - China	69.245	1.13	32	1	1	3,371,027		58.2
3	China	1	Y - Russia	215.176	2.07	10	1	1	3,712,500		57.8
4	India	0	Y - Russia	55.923	0.63		1	1	4,207,250		54.5
5	France	0	Y - NATO	55.745	2.23	27	1	1	387,635	127	63.9
6	U.K.	1	Y - NATO	48.253	1.7	15	1	0	232,675	71	78
7	Japan	0	Y - USA	46.126	3.28	2	0	4/7	311,875	305	72.3
8	Turkey	0	Y - NATO	14.803	*.45	55	0	0	743,415		65.4
9	Germany	0	Y - NATO	41.067	2.88	16	0	1	210,000	301	74.2
10	Egypt	0	Y - USA	4.513	0.72		0	0	1,329,250		53.4
11	Italy	0	Y - NATO	27.934	1.33	34	0	1	267,500	160	62.5
12	S. Korea	1	Y - USA	36.777	4.23	11	0	2/7	5,829,750	531	73.8
13	Pakistan	0	Y - USA	10.063	0.25		1	6/7	919,000	190	54.4
14	Indonesia	0	Y - USA	8.183		65	0	0	975,750		64.2
15	Israel	1	Y - USA	17.977	4.27	40	1	4/7	718,250		72.2
16	Vietnam	0	Y - USA	5.017		8	0	0	5,488,500		53.1
17	Brazil	0	Y - USA	23.676		66	0	1	1,987,000		51.4
18	Taiwan	0	Y - China	9.924			0	0	1,932,500		76.6
19	Poland	0	Y - USA	9.341	1	22	0	1	184,650		68.5
20	Thailand	0	Y - USA	5.88	0.63	57	0	0	627,425		67.1

Fig. 1 Data used in testing the model's hypotheses

Killer Robots, an organization that campaigns for a ban on lethal autonomy [17]. The second factor, whether or not a state is allied with another state that possesses LAWS, is numericized in the same way—a one to indicate that alliances exist and a zero to indicate no alliances. The logic behind the inclusion of this variable is that a state is likely to provide a given technology to its closest allies in order to achieve common goals, but a state is likely to prevent the proliferation of said technology to its adversaries to maintain a strategic advantage. For the purpose of this study, an alliance is assigned if one government party lists the other as a partner in security or military endeavors [18].

Military spending, as a quantitative variable, is presented in constant U.S. 2015 dollars [19]. Theoretically, states that spend more on their militaries are the most likely to obtain cutting-edge technology. Research and development spending is standardized as a percentage of the state's 2015 GDP (the only exception is Turkey, whose most recent statistic was from 1996) [20]. Research and development spending is included as a factor based on the assumption that lethal autonomy is a complex technology that requires a high level of development from several sectors. In addition, it remains an experimental technology in all but a few cases, and therefore, its development is still a gamble for any military.

The next variable, science test performance ranking, is taken from *Business Insider*'s 2016 ranking of participating states based upon the national average of that country's performance on a science- and math-based standardized test [21]. The goal of the test is to determine which states have the best science and math education as a way of predicting future innovation dominance. This variable is believed to correlate somewhat with a country's dependence on research and development spending as the production of highly technical software theoretically requires a highly-educated populace.

The next variable—possession of nuclear weapons—is numericized in accordance with earlier qualitative variables: one is assigned to indicate possession and zero to indicate a lack thereof. The logic behind choosing such a variable is that nuclear weapons signify a desire to attain military dominance and a credible threat of the use of force. The acquisition of lethal autonomy indicates a similar willingness, especially when one considers the previously-discussed argument that lethal autonomy is not capable of complying with international humanitarian law in conflict. Additionally, if one assumes that both weapon types drastically change the status quo of modern warfare and acquisition indicates a state's desire or willingness to do so.

As a measure of a state's engagement in international arms control efforts, a state's signatory status to the Convention on Certain Conventional Weapons (CCW) is assessed. The variable is coded by assigning it a fraction out of seven based upon the six protocols and one initial article of the treaty. The CCW is a treaty designed to limit the use of certain inhumane weapons, such as landmines and cluster munitions, and is the only treaty convention in existence which addresses lethal autonomy in weapons with a Group of Governmental Experts focused on studying and potentially restricting this technology [22]. Thus, the hypothesis for this research study is that the level of a state's engagement in the CCW is an acceptable gauge of the state's reluctance to use a weapons system considered morally or legally reprehensible by a large portion of the international legal community.

This study also includes the level of a state's bureaucracy, which is coded based upon a ranking system from The Heritage Foundation's "2018 Index of Economic Freedom." [23]. This index generates a score for each nation-state on a scale of 100, with the highest number indicating complete economic freedom and the lowest indicating a repressive absence of freedom. The total number is an aggregate of four categories—rule of law, government size, regulatory efficiency, and open markets. This index is used because it is a recent metric and due to its broad view of economic freedom. Although the index is named to emphasize its analysis of economics, it is also applicable to this research in that it measures levels of bureaucracy, which is considered an important factor in the adoption of any new technology. The hypothesis is that individuals within the bureaucracy will object to the automatization of warfare and resist the adoption of LAWS into mainstream military functions.

The "robot density" of a given state, a variable related to the bureaucracy variable discussed above, is the eighth factor analyzed in the study. The term "robot density" refers to the number of multipurpose industrial robots per 10,000 people in manufacturing and the numbers cited in this study were taken from a 2016 presentation by the International Federation of Robotics [24]. This factor is tied closely to the presence of a bureaucracy in that the effect of a bureaucracy is diminished as the robot density of a given state rises, since general exposure to robots in the workplace is likely to make them a more accepted aspect of society and, therefore, a more legitimate addition to the armed forces of a given state. Thus, the hypothesis for this variable is that the higher the robot density in a state, the more likely the state is to proliferate LAWS.

The final variable that this study will investigate is the manpower of a given state and, more specifically, the total military personnel of the state according to Global Firepower's 2018 data [25]. The rationale behind including this variable is that the manpower of a given military may directly affect its likelihood to adopt new and advanced technologies. In the same way that a nuclear weapon greatly multiplies the threat posed by even the smallest military, lethal autonomous technology could greatly increase the reach of a military with an otherwise limited force. This technology could also allow a smaller military to engage in riskier activities with far less risk to its troop contingent. Thus, the hypothesis is that states with fewer total military personnel are more likely to proliferate.

The author originally considered whether or not a state is engaged in interstate conflict as a factor influencing proliferation based upon the author's hypothesis that states engaged in conflict are likely to seek innovative ways of ending the conflict as quickly as possible. However, due to complexity in coding, the author chose to eliminate this variable and focus on more quantitative assessments. In addition, the variable "Allied with a LAWS Possessor" was eliminated from the analysis as exhibited in Fig. 1 because all states listed in the sample have relations with a possessor state with whom they could be considered allies. With the elimination of these variables, the resulting research hypothesis model emerged and is conceptually illustrated in Fig. 2.

After defining the model, each of the eight variables was tested for statistical significance to eliminate potentially insignificant variables from the final algorithm. Several tests were performed to determine how much variation in the dependent variable could be explained by the independent variable, as well as how statistically significant the correlation is between any given independent variable and the dependent variable. Figure 3 shows an example result of the analysis probing the relationship between a state's military spending and LAWS possession.

In the graph, the regression line shows a positive correlation between military spending and possession of lethal autonomous technology. In other words, the more a state spends on its military, the more likely it is to possess lethal autonomous weapons. However, the R^2 value for this regression is 0.245, indicating that military spending only explains a small portion of the change in the independent variable, LAWS proliferation. Further tests reveal that the P Value of the relationship between military spending and possession is low, with a value of 0.026. This P Value indicates high confidence in a positive correlation between the two variables. For this reason, military spending remains a statistically significant variable and is included in the final algorithm.

A negative correlation between research and development (R&D) spending and possession of LAWS is found, indicating the states who spend less on research and development are more likely to possess lethal autonomous technology, as shown in Fig. 4 This result is surprising given the technical difficulties of producing a viable autonomous system. This result also disproved the original hypothesis that R&D spending would be the most important factor in predicting possession of LAWS and may indicate the necessity of a new approach to lethal autonomy. Lethal autonomy could shift from being a technological advance to being a low-budget alternative

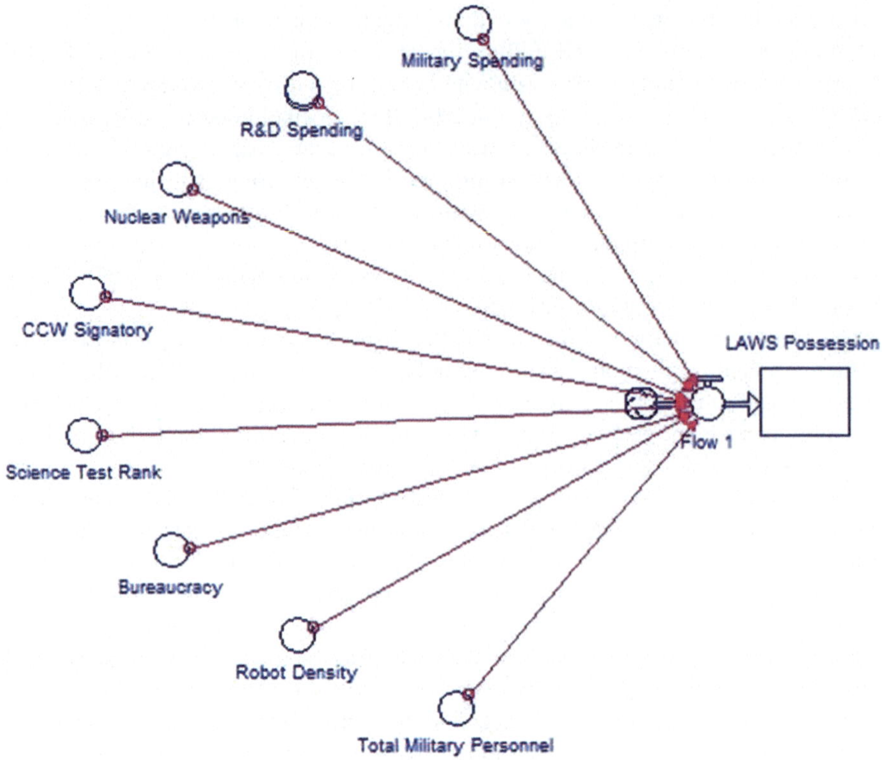

Fig. 2 Model of LAWS proliferation variables from hypothesis

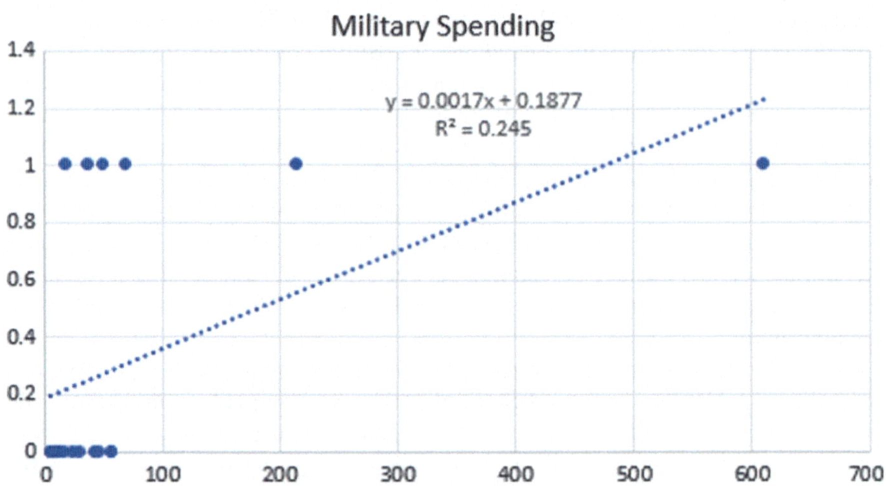

Fig. 3 Relationship between military spending and LAWS possession

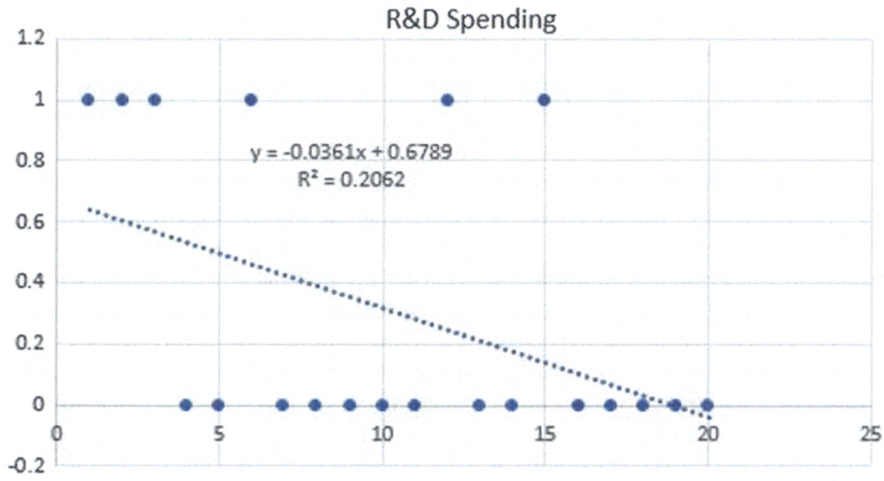

Fig. 4 Relationship between R&D spending and LAWS possession

to high-cost items, or to troops to whom pay, medical service, and housing are due in an effort to reduce overall military spending. Despite the deviation from the hypothesized result, the variable will be included in the final algorithm given that there is 95.9% confidence in the existence of a relationship between R&D spending and LAWS possession.

An analysis of the relationship between nuclear weapons possession and LAWS possession indicates a positive correlation between the two variables and results in the strongest explanatory power of any variable in the model, as shown in Fig. 5.

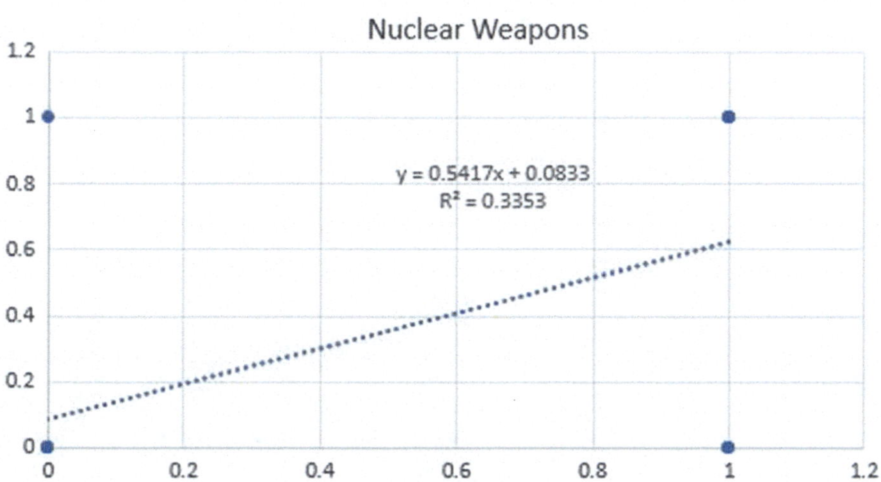

Fig. 5 Relationship between nuclear weapons possession and LAWS possession

Additionally, the P value of the regression analysis reveals 99.3% confidence in the existence of a relationship with an explanatory power of 33.5%. This makes the possession of nuclear weapons a strong indicator of whether or not a state possesses lethal autonomy, and therefore this data can be considered a strong independent indicator of possession of lethal autonomy.

The regression line for the graph of the variable, signatory status to the CCW, shows a negative correlation between the two variables and is shown in Fig. 6. A state that is a signatory to more parts of the CCW is less likely to possess lethal autonomy. Despite the agreement with the author's earlier predictions, the explanatory power of this variable (0.3%) is much smaller than that of previous variables. Additionally, a 28.1% confidence in the existence of a relationship between the two variables indicates a relationship too weak to be considered significant. Thus, a state's signatory status to the Convention on Certain Conventional Weapons cannot be independently considered a good indicator of possession, based upon the results of this study.

A probe of the relationship between average scores on science tests, as an indicator or scientific and technical education status, and LAWS in shown in Fig. 7. As the regression line of this graph shows, the negative slope indicates that the lower a state's average score on the standardized test, the more likely a state is to possess lethal autonomous technology. This relationship is another surprising find, given assumptions described earlier—specifically, the technical nature of lethal autonomy and the need for an educated populace.

The phenomenon of low science test scores in the United States despite a highly active technology industry and clear demand for technological innovation has been previously written about [26]. The United States is a "brain drain" benefactor for foreign scientists as individuals educated abroad through the primary level immigrate to the United States for secondary education or on visas for full-time employment.

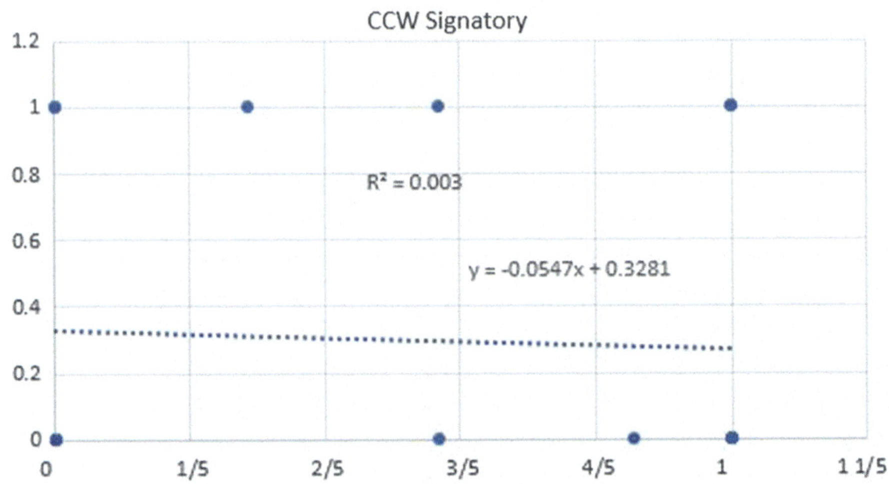

Fig. 6 Relationship between CCW signatory status and LAWS possession

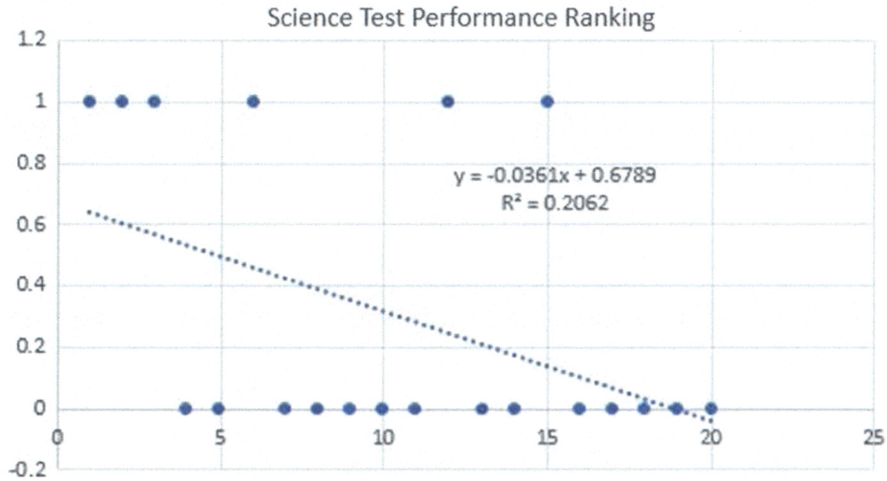

Fig. 7 Relationship between science test performance ranking and LAWS possession

In this way, the US can afford low test scores without falling behind on technological innovation. Although this explanation can account for the trend exhibited in the graph, it does have limits as how the phenomenon plays out in other countries is not well studied. Additionally, the negative slope of the graph aligns with the findings of the research and development graph presented earlier. A P-value indicating 76% confidence in the existence of a relationship is not high enough to be a statistically significant indicator of LAWS possession.

The analysis of the relationship between a state's level of bureaucracy and LAWS possession is shown graphically in Fig. 8. It is found that the more bureaucratic a state is corresponds with greater likelihood that a state possesses lethal autonomous technology. This is a surprising result that is contrary to the original hypothesis that a bureaucracy would impede proliferation of a controversial new technology.

In the case of the U.S., the positive correlation between the two variables could be explained by the introduction of the National Defense Authorization for fiscal year 2001 in which the US Congress called for one third of all operation ground combat vehicles to be unmanned by 2015 [27]. One of the bureaucratic creations that should, according to the hypothesis, discourage the adoption of lethal autonomy into the military infrastructure may conversely being driving the opposite result. Regardless, this variable carries a P value of 0.148, indicating just 85% confidence in the existence of a relationship between level of bureaucracy and possession of LAWS. Additionally, the R-squared value for this variable indicates that only 11% of the variation in possession can be explained by variation in level of bureaucracy; thus, this variable is not independently a good measure of likelihood of LAWS possession.

The positive slope of the regression line Fig. 9 supports the original hypothesis—that a state with a higher robot density is more likely to possess LAWS. The P value

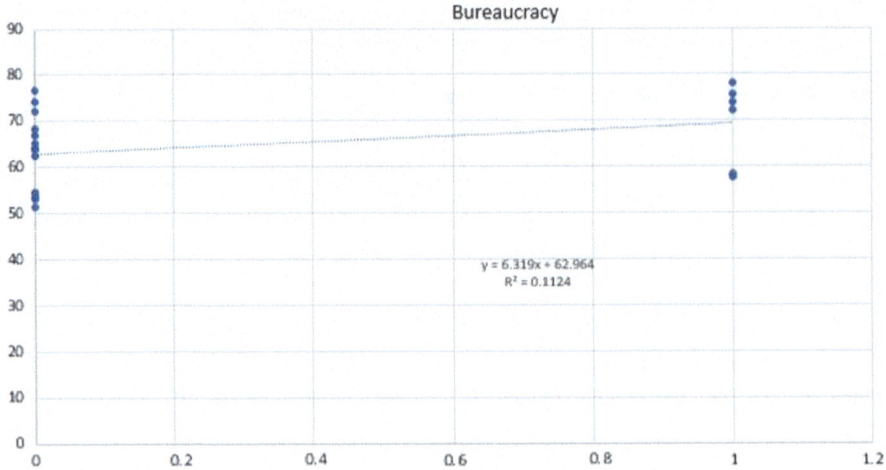

Fig. 8 Relationship between level of bureaucracy and LAWS possession

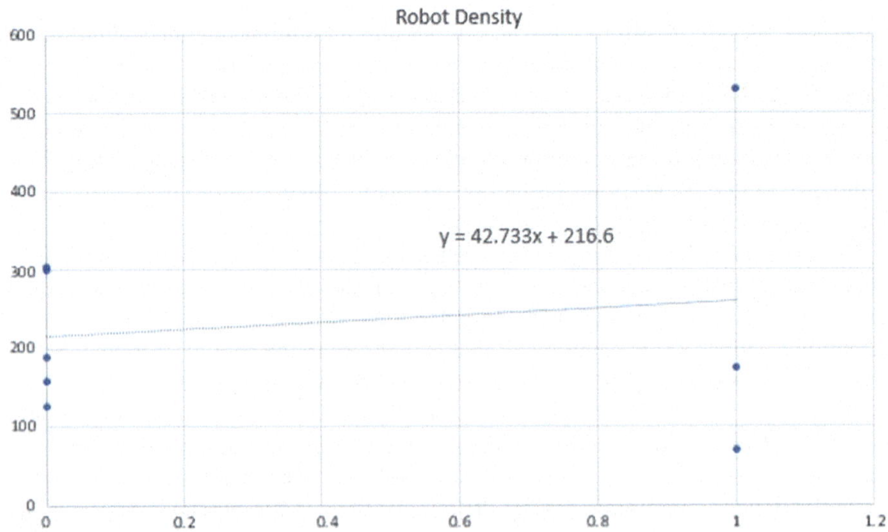

Fig. 9 Relationship between robot density and LAWS possession

for this variable is much higher than that of other variables within this data set and reveals that one can only be 28% confident that a relationship exists between robot density and LAWS possession. Furthermore, an R-squared value indicates that only 2.3% of the variation in possession can be explained by LAWS. This result makes this variable an extremely poor predictor for LAWS possession—at least when it is not combined with other variables. This variable could also have limited strength due to a lack of data. While other variable analyses incorporated data for nearly all

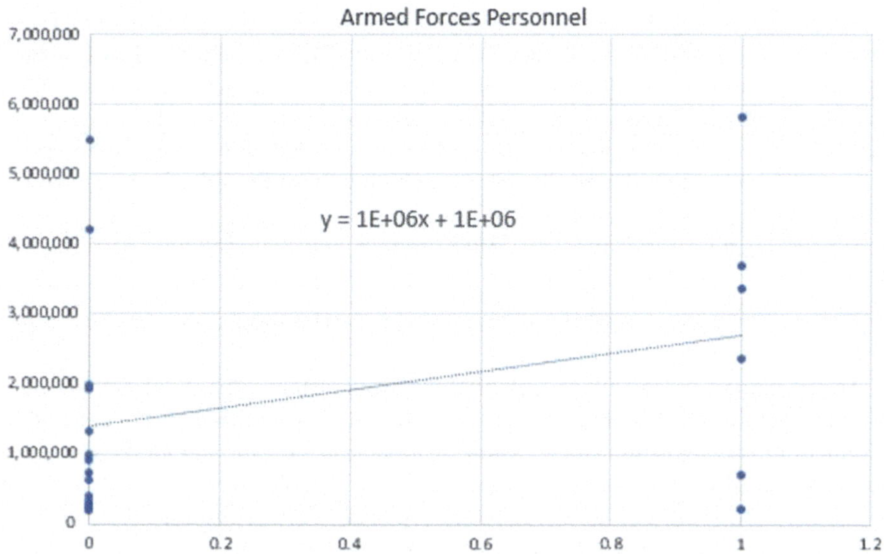

Fig. 10 Relationship between number of armed forces personnel and LAWS possession

of the twenty states under scrutiny, the lack of available data limited this particular analysis to only a few data points.

The relationship between the number of people in a nations military and LAWS possession is shown in Fig. 10. The slope of this graph indicates a positive correlation between the total number of military personnel and possession of lethal autonomous weapons systems. This finding runs contrary to the original hypothesis, which posited that the adoption of lethal autonomy into a military could boost its reach, making LAWS an attractive option for smaller militaries. The opposite argument, however, is that larger militaries may be more capable of expanding into riskier and more experimental technologies than militaries with more limited resources, making larger militaries a more likely candidate for possession. The P value for this variable indicates 86% confidence that a relationship exists between the two variables, while the R-squared value shows that 12% of the variation in possession of lethal autonomy may be explained by the number of total military personnel that a state possesses. Thus, this variable may be reasonably important in predicting proliferation and may form an integral part of the algorithm.

5 Deriving the Relationship

The statistical tests on the eight variables determine the significance as reliable indicators individually and demonstrate each variable's relationship to the dependent variable. Thus, each variable can be assessed as a single entity. While this method is useful in determining the individual utility of each variable as an indicative measure or warning sign of proliferation, the tests fail to form a bigger picture. Thus, combining the variables into one algorithm will enable a more comprehensive analysis of the importance of each variable relative to the others and will reveal whether or not other variables are necessary to predict the proliferation of lethal autonomous weapons systems. The algorithm that results from combining these variables is shown below.

$$y = 0(Military\ Spending) - 0.289(Research\ and\ Development\ Spending) \\ -0.013(Science\ Test\ Ranking) + 0(Total\ Military\ personnel) \\ -0.002(Robot\ Density) + 0.045(Bureaucracy) - 1.76$$

This formula is derived from calculations using ordinary least squares and uses coefficients to weight the importance of each variable. Two variables are noticeably absent from the algorithm—possession of a nuclear weapon and CCW signatory status. When calculating the algorithm, these two variables were eliminated due to collinearity, meaning that their explanatory power overlaps with another variable and, therefore, the two eliminated variables became redundant.

Despite earlier predictions of the importance of military spending as a variable, as well as relatively strong confidence in the existence of a relationship, comparatively the relationship was too small to be significant in the algorithm. Although the coefficient for military spending is not zero, the coefficient is so small that rounding to the nearest thousandth still eliminated the number. This could be the result of a possible outlier in military spending: the United States possesses LAWS and spends significantly more on its military than any of its contemporaries at the top of the Global Firepower ranking. The same phenomenon occurs with total military personnel. Therefore, the revised algorithm is as follows:

$$y = -0.289(Research\ and\ Development\ Spending) \\ -0.013(Science\ Test\ Ranking) - 0.002(Robot\ density) \\ +0.045(Bureaucracy) - 1.76$$

This revised algorithm is designed to predict possession of lethal autonomous weapons systems, or y. The coefficients for science test ranking and robot density are both negative.

In theory, a user could insert values for each independent variable and calculate a number between zero and one that would indicate how likely a state is to possess lethal autonomy—a zero indicating extremely unlikely and a one indicating a high risk of current or future possession. If this method were applied to a list of states, it would be possible to determine which states were the most likely to develop or obtain these weapons systems in the future. In doing so, the user could determine the most appropriate target states for further monitoring.

Despite the elimination of military spending and total military personnel as unnecessary variables, the algorithm retains several critical problems. The very nature of the investigation renders the mathematics of the equation extremely difficult if not impossible. This difficulty stems from the high degree of relationship between the variables, which affect one another and therefore violate the principles of basic statistical assumptions. For this reason, an R squared value for the equation is impossible to obtain, as are many of the other benchmarks for determining the accuracy and efficacy of a given set of variables. Other problems arise due to an insufficient sample size, an insufficient variable count, or a combination of these issues.

Nonetheless, the results revealed important findings both directly regarding factors that correlate with autonomous lethal weapons and more broadly on the need for more robust scholarship, both qualitative and quantitative, to understand the impact of disruptive and emerging technologies on national security.

6 Discussion of the Results

This research began by seeking the answer to three questions regarding the proliferation of lethal autonomous weapons systems: What are the key factors in determining whether or not a state will develop lethal autonomous weapons systems? Can these factors be used to predict the proliferation of this technology? If so, which states are most likely to obtain LAWS in the future? The analyses of individual research reveal that, contrary to the original hypothesis, research and development spending is not the strongest indicator of LAWS possession. Instead, the possession of nuclear weapons is the most statistically significant variable in predicting LAWS possession. Military spending was the second most-useful variable in predicting LAWS possession, followed by research and development (R&D) spending. According to the individual analysis, science test ranking and CCW signatory status had the least-certain relationship in influencing LAWS possession.

This research study sought to create an algorithm capable of predicting a country's likelihood of LAWS possession. Despite the study's success in determining the importance of each variable individually in relation to LAWS possession, the combination of these variables into an algorithm did not create a cohesive image of proliferation patterns with full explanatory power. The initial algorithm revealed that military spending had a miniscule relationship to possession and, therefore, it was dismissed as an element of the algorithm. The results of the final algorithm were consistent with individual-variable assessments' results maintaining that nuclear weapons are the most useful indicator of a state's likelihood of possessing lethal autonomous weapons.

Although the creation of an algorithm was a success in that it identifies several important variables in relation to LAWS proliferation, the algorithm itself lacks predictive capabilities due to a high error value. Additionally, the algorithm cannot explain 35.5% of variation in states that possess LAWS. Thus, the second question this study seeks to answer concerning whether factors can be used to predict the proliferation of this technology must be answered in the negative, within the constraints of this research study. Since the third and final question concerning which states are most likely to obtain LAWS in the future is dependent upon a positive answer to the second question, it cannot be answered by the findings contained herein. The approach to designing a predictive algorithm within this study has several potential flaws which fall into two categories—methodological errors and conceptual errors—both of which will be enumerated upon in this section.

6.1 Methodological Errors in the Algorithm

As previously mentioned, the error value in the algorithm may be attributed to two causes. One of these is the insufficiency of the sample size, which could be increased by looking at a larger number of states. However, increasing the sample size could be difficult given that the Campaign to Stop Killer Robots lists twelve states as current possessors, but names only six of them specifically [17]. Thus, further research would be necessary to pinpoint the remaining six possessors of LAWS technology. The second limitation of this study may be an insufficient variable count, which could be remedied through further research and analysis of other variables that could be related to a country's choice to pursue LAWS technology. Adding these variables to the algorithm would reduce the number of unrecognized variables that affect possession, thereby diminishing the variation in LAWS possession not currently explained by the algorithm.

6.2 Conceptual Errors in the Algorithm

The algorithm, while hypothetically useful in a defined state, lacks several important components. In an effort to rely on statistical analyses and quantitative data, the author excluded several variables based upon individual criteria despite the fact that these variables could have an important effect on proliferation when combined with other variables. Among these concerns is the fact that the algorithm fails to consider non-state actors as potential beneficiaries of proliferation. Assuming that this study's results are accurate and states with low research and development spending are more likely to possess lethal autonomous systems, non-state actors are candidates for possessing such technology. The utility of such a system would be significant for a terrorist group; al-Qaeda for instance struggles to maintain a membership high enough to carry out an effective campaign, but with LAWS technology, a non-state actor could hypothetically execute the missions without the loss of life associated with suicide bombers.

Furthermore, the factors that currently restrain states from using lethal autonomy in combat situations may not be as important to non-state actors. As previously mentioned, discrimination in a conflict zone is critical to compliance with international law, but international law holds only states accountable and is not applicable to non-state actors. The ability of an autonomous weapons system to discriminate between targets is not important when the goal of use is to spread fear. Thus, the algorithm's failure to apply to non-state actors is a serious gap in its utility, while correcting this gap would require an entirely new set of variables for analysis.

6.3 Future Research and Expansion

Given the limitations in the research design due to paucity of information available on LAWS possessor states and the conceptual gaps in this research, further improvements in the research outlined in this study are warranted. While the opportunities for improvement are numerous, the author considers the next expansion stage in this research as twofold. The preliminary step involves improving the research model, which is in itself a multi-step process. Initially, the model requires a larger sample size to increase confidence in the results. Having a larger sample size, however, necessitates identifying more states that currently possess lethal autonomous weapons technology. Once the sample size is expanded, researchers could identify additional variables to test. As the sample size and variable counts become larger, the algorithm will increase in legitimacy as a predictive algorithm and gain more explanatory power. Theoretically, once this change occurs, the algorithm might be capable of predicting a state's LAWS possession, and researchers would be able to determine which states are most likely to obtain LAWS in the future.

The second step in expanding upon the current research framework might be to extend the algorithm beyond an international-level analysis to include non-state actors. Budgetary variables for non-state actors could replace the variables such as research and development spending in the assessment of state actors. As no non-state actor is currently known to possess nuclear weapons, that variable might be replaced with another variable, such as evidence of a non-state actor seeking nuclear materials. Additionally, for non-state actors, motivation and population may become more important factors. The main impediment with testing an algorithm designed to monitor proliferation to terrorist groups is difficulty in gathering information, as terrorist groups lack the same compulsion for transparency that the international community requires of states. Thus, this suggested research path is an example of research that would benefit substantively from more qualitative, rather than quantitative, methodologies. Private organizations could also find utility in autonomous robotics; they could be used in monitoring environmental conditions or securing private property from intruders. The difference in this use is that, to monitor environmental conditions (in the Arctic, for instance), the technology does not require lethal capabilities and, therefore is outside of the scope of this research.

The third step in expanding upon this foundational research is creating a more in-depth analysis of opinion regarding the proliferation of lethal autonomy. While this research touches on this topic briefly in earlier sections, a more thorough investigation of the origins of public opinions could be useful in determining the root of the push for proliferation. Research in this area could be expanded by examining statements about and literature on LAWS from several different groups within society—academia, philosophy, military, and government—to analyze conflict within and among each of the four societal clusters. In doing so, one could hypothetically locate several key points of origin for the push for proliferation, indicating the driving force for the conversation around lethal autonomy.

7 Conclusions

The proliferation of any weapons system, specifically one as versatile as a lethal autonomous weapons system, is difficult to predict given the volatility of international relations and the fluidity of conflict. Ultimately, this research failed to answer two of the three key research questions. However, the results were conclusive in determining which of the analyzed variables are the most critical indicators of possession of LAWS and did result in the creation of a model. Although the current algorithm is imperfect and lacks predictive capability, the research outlined in this chapter lays a foundation for future expansion and is indicative of research necessary in the broader realm of emerging and disruptive technologies. Furthermore, the existing model does show promise for future results, as statistical analysis shows explanatory power for 64.5% of the variation in possession of LAWS. As outlined in the previous section, many exciting possibilities for future research exist, with expanded application and more variables to increase the algorithm's predictive capability. Just as important, the current research framework did reject several variables and narrowed the field of potentially relevant factors in predicting proliferation. Despite the model's inability to make concrete predictions, the research was successful in several aspects and is effective in its overall goal of expanding the knowledge base on this groundbreaking new military technology.

References

1. Marchant GE et al (2011) International governance of autonomous military robots. Columbia Sci Technol Law Rev 272
2. Purkiss J, Serle J (2017) Obama's covert drone war in numbers: ten times more strikes than bush. Bureau Investig Journal. https://www.thebureauinvestigates.com/stories/2017-01-17/obamas-covert-drone-war-in-numbers-ten-times-more-strikes-than-bush
3. The Threat of Fully Autonomous Weapons (2018) Campaign to Stop Killer Robots. www.stopkillerrobots.org/learn/
4. Navy installations underway with transitioning to defense biometric identification system to enhance base security. United States Navy, 21 Apr 2017. https://www.navy.mil/submit/display.asp?story_id=100036
5. TALON: Medium-Sized Tactical Robot (2018) QinetiQ. https://qinetiq-na.com/products/unmanned-systems/talon/
6. BAE Systems to Lead £124 Million UAV Technology Demonstrator Programme. defenseaerospace.com, 7 Dec 2006. http://www.defense-aerospace.com/articles-view/release/3/76514/industry-team-details-role-in-british-ucav-program.html
7. Taranis (2019) BAE Systems. http://www.baesystems.com/en/product/taranis

8. Kalashnikov Gunmaker Develops Combat Module Based on Artificial Intelligence. TASS, 5 July 2017. http://tass.com/defense/954894
9. Thurnher JS (2014) Examining autonomous weapons systems from a law of armed conflict perspective. In: McLaughlin R, Hitoshi N (eds) New technologies and the law of armed conflict. Asser Press, The Hague
10. Unmanned Systems Roadmap: 2007–2032. US Department of Defense, 10 Dec 2007, p 49. https://www.hsdl.org/?abstract&did=481851
11. Arkin Ronald C (2010) The case for ethical autonomy in unmanned systems. J Mil Ethics 9(4):332–341
12. Sharkey NE (2012) The evitability of autonomous robot warfare. Int Rev Red Cross 94(886), pp 787–799. http://e-brief.icrc.org/wp-content/uploads/2016/09/23.-The-evitability-of-autonomous-robot-warfare.pdf
13. Joint Concept for Robotic and Autonomous Systems (JCRAS). Joint Chiefs of Staff, US Department of Defense, Oct 2016
14. Hughes WP (2015) Robot ethics and future war. In: Consortium for Robotics and Unmanned Systems Education and Research (CRUSER) News, vol 5
15. Meyer SM (1986) Dynamics of nuclear proliferation. University of Chicago Press, Chicago, pp 160–162
16. 2017 Military Strengh Ranking (2017) Global Firepower. http://www.globalfirepower.com/countries-listing.asp
17. The Problem (2018) Campaign to Stop Killer Robots. https://www.stopkillerrobots.org/learn/#problem
18. What is NATO? North Atlantic Treaty Organization. http://www.nato.int/nato-welcome/index.html#members
19. SIPRI Military Expenditure Database (2018) Stockholm International Peace Research Institute (SIPRI). https://www.sipri.org/databases/milex
20. Research and development expenditure (% of GDP). World Bank. https://data.worldbank.org/indicator/GB.XPD.RSDV.GD.ZS?locations=FR
21. Jackson A, Kiersz A (2016) The latest ranking of top countries in math, reading, and science is out—and the US didn't crack the top 10. Business Insider, 6 Dec 2016. www.businessinsider.com/pisa-worldwide-ranking-of-math-science-reading-skills-2016-12
22. Convention on prohibitions or restrictions on the use of certain conventional weapons which may be deemed to be excessively injurious or to have indiscriminate efworld robotics report-fects. In: Aka the convention on certain conventional weapons. The United Nations Office at Geneva. http://disarmament.un.org/treaties/t/ccwc
23. Country Rankings. The Heritage Foundation. http://www.heritage.org/index/ranking
24. World Robotics Report 2016. IFR International Federation of Robotics, 29 Sept 2016. https://ifr.org/img/uploads/Presentation_market_overviewWorld_Robotics_29_9_2016.pdf
25. 2018 Military Strength. Global Firepower. http://www.globalfirepower.com/country-military-strength-detail.asp
26. Paarlberg RL (2004) Knowledge as power: science, military dominance, and U.S. security. Int Secur 29(1):138–145
27. National Defense Authorization, Fiscal Year 2001. In: 106th Congress, Washington, D.C., vol 40. U.S. Government Publishing Office, 30 Oct 2000. https://www.govinfo.gov/content/pkg/PLAW-106publ398/pdf/PLAW-106publ398.pdf. Accessed 6 Feb 2019

Protecting Army Aviation and Enabling Military Dominance Through Disruptive Innovation

T. Christopher Petty

Abstract State and non-state actors are challenging the US Army's asymmetric strategic advantage and global military dominance. Countries such as Russia and China are incorporating technological advances to limit and challenge the US Army's capabilities and freedom of maneuver. Non-state actors are acquiring advanced weaponry creating a hostile environment for Army aviation operations in the Middle East, and these revisionist elements challenge the US's hegemony making the world more violent and unstable. To thwart the projected actions of state and non-state entities, the US Army needs to undergo another disruptive innovation in military power. The Persian Gulf War was a demonstration of the US military's last disruptive innovation resulting in global military dominance, and the US Army is currently planning for the next leap in military capability. This paper proposes that the path to the next disruptive innovation is through investments in the foundational technologies of nanotechnology, quantum computing, and artificial intelligence. Applying these innovations to Army Aviation Advanced Protection Systems creates an asymmetric strategic advantage ensuring global freedom of maneuver for the US Army's land forces.

1 Introduction

> Learning and innovation go hand in hand. The arrogance of success is to think that what you did yesterday will be sufficient for tomorrow. William Pollard [1]

On February 24, 1991, United States (US) forces along with a coalition force from 39 other nations, attacked across the Saudi Arabian/Iraqi border to liberate Kuwait. The US Army, which deployed in 1990 to Saudi Arabia, the product of almost twenty years of reform and experimentation, bore little resemblance to the Army that left the Republic of Viet Nam in 1972 [2]. It was a technologically superior force unlike any

T. C. Petty (✉)
US Army (Retired), Seattle, WA, USA
e-mail: tcpetty58d@gmail.com

M. E. Kosal (ed.), *Disruptive and Game Changing Technologies in Modern Warfare*, Advanced Sciences and Technologies for Security Applications,
https://doi.org/10.1007/978-3-030-28342-1_10

seen in previous conflicts. The US underwent an offset strategy in the 1970s, which helped to create the revolutionary new systems such as extended-range precision-guided munitions, stealth aircraft, advanced rotary wing aircraft, and ground vehicles that created the dominating military force unleashed during the Persian Gulf War [3]. This overwhelming military power application and lessons from the Persian Gulf War did not go unnoticed. It served as a wake-up call for nations like Russia and China highlighting how much of a conventional military capability gap existed between them and the US. The Persian Gulf War served as their catalyst to bridge the gap and challenge the US global hegemony and military dominance. Technologies that were rare and sophisticated in 1991 are now commonplace. Russia and China played catch-up with costly modernization programs while also seeking asymmetric counters to US conventional advantages [4]. Since 1991, state actors such as Russia and China and non-state actors like Al Qaeda and Islamic State (Daesh) have been innovative to reduce and match US Army's capabilities displayed during the Persian Gulf War and the Global War on Terrorism [5].

In response to the changing global power dynamic, in 2014 the US Department of Defense (DOD) began an offset strategy utilizing a disruptive, innovative technology called the Defense Innovation Initiative (DII). The DII is an ambitious Department-wide effort to identify and invest in innovation to sustain and advance America's military dominance for the 21st century [3]. The DII is the US Army's next growth in power, its next "Major Military Innovation." According to Michael Horowitz, Major Military Innovations (MII) are significant changes in the conduct of warfare, relevant to leading military organizations, designed to increase the efficiency of which capabilities convert to power [6]. It is this military power the US Army and the DOD are trying to harness through technological advancements to increase military capability. Though no longer utilized for planning and shaping the force, the US Army is on the cusp of a new Revolution in Military Affairs (RMA) [7, p. 38]. Whatever terms the US Army chooses, it still needs an RMA-like evolution to catapult the US Army's military dominance beyond any near-peer state or a non-state actor. History has shown those slow to adapt to military revolutions and revolutions in military affairs are likely to suffer painful results. When the pace of change accelerates, the militaries that anticipate and adapt are likely to gain a massive advantage over potential enemies who are less agile [7, p. 39]. The US Army may not call it an RMA, but it is planning, preparing, and investing for future warfare against a breadth of potential adversaries, and like the Persian Gulf War, technology will play a crucial role.

This study analyzes what technologies are required and why, in a resource-constrained environment, the US Army must invest into deliver continued military dominance by focusing on enhanced protection capabilities in Army Aviation platforms. Additionally, this study addresses how those innovations will provide the asymmetric strategic advantage the US Army requires to execute its missions successfully. By investing in nanotechnology, quantum computing, and Artificial Intelligence (AI), the Army can make a resourced, informed investment strategy to achieve continued military dominance. This strategy creates the environment to develop disruptive technologies for Future Vertical Lift (FVL)/legacy aircraft Advanced Protection System (ADPROS) (Air) capability. This enhanced capability provides the US Army with the necessary RMA to achieve asymmetric dominance needed to deter and deny a state or non-state adversary strategically, and ensure freedom of maneuver for a Joint Force operating globally.

To support my recommendations, I will describe the future operating environment the US Army expects to fight within and the actions it is implementing to maintain dominance as a land based maneuver force. Then I will outline Army Aviation requirements to enhance ADPROS capabilities and make recommendations for US Army investments in nanotechnology, AI, and quantum computing to produce an unmatched asymmetric land force. I will provide an overview of my quantitative method used to support my argument and close with recommendations for future studies and actions for future innovative development investment.

2 The Future-Operating Environment

> No matter how clearly one thinks, it is impossible to anticipate precisely the character of future conflict. Sir Michael Howard [8, p. 4]

Sir Michael Howard described the chaos and the essence of war through his studies and personal military experiences. He also related the profound impacts of technology and how it changed warfare throughout time [9]. Understanding the impacts of technology married with knowledge of potential adversaries allows the US Army to determine what capabilities it needs. An example is the US Army's focus on the specific technologies required to protect its aircraft that execute military operations globally against threats from state to non-state actors. It is important to understand how the US Army sees the global threat in the future Operating Environment (OE), specifically the threat against the US Army's multiple air systems. Understanding these threats will highlight why technological innovation is critical to not only survivability, but also continued US military dominance. To maintain this dominance, the US Army must determine what the world will be like in the next thirty to fifty years. The US Army's forecast does not have to be perfect, but it is important to get as close to right as possible.

The US Army basis for threat analysis assumes all threats originate from land. These land-based threats can come from state or non-state actors operating within the land, air, maritime, space, and cyberspace domains [8, p. 9]. To plan appropriately, the US Army established a planning foundation to drive how it will fight, train, and resource to maintain a dominant military position in the world. The US Army focused considerable effort to define the future operating environment so that the doctrine, capabilities, and systems can match and dominate the projected threats.

The future operational environment (OE) will be complex, dynamic, and shaped by the convergence of a myriad of global trends driven by social dynamics, nature, and technology. Pervasive characteristics of the OE include increased velocity of human interaction and momentum of events; potential for overmatch; proliferation of weapons of mass destruction; spread of advanced cyberspace and counter-space capabilities; ease of technology transfer to state, non-state, and hybrid actors; transparency and ubiquitous media; demographics, and operations among populations in megacities, dense urban areas, and in complex terrain [10, p. 4].

The threats the US Army will face in the future are complex, diverse, and ever-evolving. These emerging threats may employ the full range of military capabilities including traditional conventional operations, unconventional, hybrid, and criminal activities to target vulnerabilities and negate the US Army's technological advantages [10, p. 8]. Our nations' enemies will acquire and leverage technology to close the capability gap and erode US Military dominance in all domains. Potential adversaries will leverage emerging technologies to disrupt the US Army's advantage in communications, long-range fire, and surveillance [8, p. 10].

We do not have to wait until 2025 to see what the global threat is like; we are starting to get a snapshot into the future OE from the threat indicators we have now. In Europe, Russia is expanding territory with the annexation of Crimea and asserting its power over neighboring states through strategic Diplomatic, Information, Military, and Economic (DIME) actions. China is undergoing a massive military modernization program, creating regional tension through territorial disputes to expand their influence. Also in Asia, North Korea continues to develop and refine a nuclear and long-range missile capability. In the Middle East, Iran continues to support extremist groups to undermine US policies across the globe, and the Islamic State of Iraq and Syria (ISIS) still poses a threat to western society globally [11]. All of these opposing efforts are to marginalize US military dominance to further state and non-state objectives, which add to the potential global chaotic instability expected to 2025 and beyond. Global instability challenges the US military's strategic objectives and provides numerous opportunities for state and non-state actors to redefine the global society to suit their agendas. The state actors such as Russia and China, are investing to improve and modernize their militaries.

These potential adversaries are modernizing their militaries with the aim of countering the US's ability to project military force. These nations and others are developing Anti-Access/Area-Denial (A2/AD) strategies such as advanced air defense systems, long-range artillery, rockets, missiles, cyber-attacks and/or electromagnetic warfare [12, p. 3].

Non-state actors leverage the chaos to access conventional weapon systems, such as shoulder-fired anti-aircraft missiles, posing a direct threat to US military aircraft as they operate globally [13]. The greatest future threat to the continued US Army military dominance is revisionist state actor advancements in A2/AD systems and the proliferation of surface to air missiles gained by non-state actors. Understanding these threats allows the US Army to plan appropriately, train, resource and win in a world that is unknown and unknowable [8, p. 3].

3 Preparing for the Future-Operating Environment

> To secure peace is to prepare for war. Carl Von Clausewitz [14]

Potential adversaries around the world are innovating, modernizing, and investing to increase their military capacity for expanding their global agendas. The pervasiveness of more lethal and technologically advanced weapons throughout the OE increases the importance for the US Army to understand the entities that may employ those systems [10, p. 5]. So what is the way ahead for the US Army? How does the US Army maintain its global military dominance to secure the peace? The US Army is working to answer these questions through multiple simultaneous and mutually supporting efforts. The first effort is understanding the future environment within which it will operate. Through understanding the environment and how potential adversaries utilize technology allows the US Army to determine the capabilities needed to defeat emerging threats. This understanding guides the other efforts for technological development giving US forces the desired asymmetric advantage to win and succeed in a multi-domain fight.

To overcome the operational challenges and emerging threats in 2025, the Army must develop future capabilities, including the ability to operate freely in the electromagnetic spectrum, maintaining secure and reliable communications as well as accurate position, navigation, and timing (PNT) capabilities [10, p. 5]. The US Army must develop advanced protection and defense platforms to defeat progressively more technologically advanced threat protective systems. The Army must be prepared to advance the capabilities and employment of directed energy weapons along with enhanced conventional capabilities. Future Army forces will project power by applying cross-domain capabilities from land to create synergy across all domains, ensuring Joint Force freedom of movement and action. In addition to working throughout multiple domains, the Army will have to develop effective capabilities to protect friendly forces, information, and systems, detect adversary threats, react to indications and warnings, and restore capabilities when challenged by adversary systems or tactics [10, p. 6].

Understanding the future OE directly shapes the US Army's investment strategy and enables the US Army to develop its capability needs to drive technological requirements. However, this is only part of the US Army solution. The US Army simultaneously developed new doctrine and leader development programs to mutually support the required technological advances creating a solid foundation for the US Army's future force. Since 2002, the US Army has focused on fighting the wars in Iraq and Afghanistan. The technological innovations and investments made by the US Army focused on supporting global counter-insurgency combat operations. This emphasis limited modernization for future fights against the more technologically advanced enemy [10, p. 7].

To alter this course, the US Army set itself in a new direction, which requires preparation for future conflicts. In May 2012, the Army released US Army Doctrine Reference Publication (ADRP) 6.0, Mission Command [15]. The ADRP 6.0

served as a philosophical shift in how leaders execute combat operations. By training and empowering leaders through mission command, it enables disciplined initiative within the commander's intent to empower agile and adaptive leaders in the conduct of unified land operations [15, p. 1]. The next building block for leadership development doctrine was the US Army's Training and Doctrine Command's (TRADOC) Pamphlet 525-3-1, The U.S. Army Operating Concept: Win in a Complex World in 2014 [8].

The Army Operating Concept (AOC) describes how future Army forces will prevent conflict, shape security environments, and win wars while operating as part of our Joint Force and working with multiple partners. The AOC guides future force development by identifying first order capabilities that the Army needs to support US policy objectives. It provides the intellectual foundation and framework for learning and for applying what we learn to future force development under Force 2025 and beyond [8, p. 1].

The final critical component the US Army is embarking upon is prioritizing and investing in science and technology investments. These investments enable future force development to counter and dominate potential future adversaries. In September 2016, the US Army Capabilities Integration Center (ARCIC) released the Warfighters' Science and Technology (S&T) Needs [10]. This document provides an overview of the US Army capabilities needed to guide and inform those who develop material for the Army. It also provides the foundation for the prioritized systems the Army forces must have to fight and win in the future [8, pp. 3, 8]. The US Army's S&T Needs document outlines in detail the technical systems and capabilities to fight and win in the future. These systems are Future Vertical Lift (FVL), Combat Vehicles, Cross Domain Fires, Advanced Protection, Expeditionary Mission Command/Cyber Electromagnetic Activities (CEMA), Robotics and Autonomous Systems (RAS), and Soldier/Team Performance Overmatch.

This study specifically focuses on the technology for advanced protection systems in relation to FVL and legacy US Army aircraft and its direct impact on the strategic goals of the United States. The reason for including legacy US Army Aircraft in this study is due to the budget-constrained environment. Due to the cost of military operations, the US Army will continue to maintain multiple aviation systems, and any newly developed protective capabilities will increase survivability for these legacy systems. The FVL is a required capability and a critical component for achieving the agility and flexibility needed to serve a pivotal role in maintaining the United States' strategic dominance in the future.

FVL enables Joint Forces Commanders to dictate the terms of operations and take the fight to the enemy at the time and place of the US' choosing, as opposed to when conditions are favorable and this allows a greater capability to overcome the constraints of complex terrain, higher altitudes, extreme temperatures, and extended distances while performing operations throughout the Joint Operations Area. The FVL family of systems will provide expanded vertical take-off and landing capabilities in the areas of range, speed, payload, reliability, high fuel efficiency, and reduced logistical footprint [10, p. 9].

The limitations for FVL systems and current legacy Army aircraft still exist because the threat the US Army will face in the future is complex and ever-changing. The solution is to enhance FVL and current legacy aircraft capabilities by applying technological research and innovation to ADPROS. Aging aircraft require advanced protection capabilities due to their age, the need for more capable forces with fewer transport and sustainment needs to meet expeditionary requirements, and the advancement of lethal threat capabilities. Without advanced protection capabilities, future Army forces lose differential advantages in aviation as well as in overmatch against near-peer threats within Brigade Combat Teams [10, pp. 10–11].

The US Army is at a critical crossroad. Faced with the dynamic and unpredictable future OE, the US Army identified gaps in its current capabilities and, through detailed analysis, determined a resourced-informed direction for S&T investments to sustain the US Army's overmatch capability. What remains to be determined is where the US Army should invest to drive the development of the systems and capacities needed in a fiscally constrained environment. The world is on the brink of a technological revolution. The speed of advancements and the disruptive impacts on society will create a more chaotic world but serve as an opportunity for the US Army to achieve the technological innovation needed to maintain military domination. The upcoming revolution will yield unlimited and unprecedented processing power, storage capacity, and access to knowledge.

The Army's investment opportunities in this rapidly advancing field include leveraging emerging technology breakthroughs in fields such as AI, robotics, the Internet of Things (IoT), autonomous vehicles, 3-D printing, nanotechnology, biotechnology, materials science, energy storage, and quantum computing [16]. The author analyzed the above technologies with the intent to determine the best investment options to develop advanced protection capabilities for US Army aircraft. The goal is to identify the best investment opportunities, which will affect multiple capabilities at the lowest cost to the Army. This study will focus on nanotechnology, quantum computing, and AI as investment strategies to achieve the needs for Army Aviation.

4 Army Aviation Protection Needs for Strategic Dominance

Army Aviation has strived to plan and prepare for the future OE by conducting a detailed analysis to validate requirements to bridge the capability gaps identified from the US Army's Warfighting Challenges [17]. Utilizing this, ARCIC and Army Aviation identified very specific advanced protection requirements critical to increasing Army Aviation's survivability for future and legacy air platforms and to maintain asymmetric advantage. This ADPROS includes universal threat detection, open architecture, and multi-tiered defeat constructs to counter a wide array of threats [12, Annex 3-2].

Army Aviation requires universal threat detection because the current threat-system development cycle is slow, expensive, and requires a specific solution for a specific missile threat [18]. This requires Army Aircraft to maintain multiple systems, which may not be interoperable, to defeat a specific threat while adding additional weight to US Army aircraft. Army Aviation needs a universal threat agnostic detection capability that can detect, track, and defeat any incoming missile regardless of seeker type or algorithm [18].

To enable a universal threat detection system, Army Aviation requires an open architecture network, meaning Army aircraft can relay threat information between airframes and between air and ground assets. The systems communicate autonomously with other onboard systems on other platforms through an integration framework and an intelligent engagement controller. The engagement controller is a software application that serves as the brain of the distributed survivability suite. It accesses all onboard survivability systems, including missile warning systems, hostile fire detection systems, laser warning receivers, radar-warning receivers, and electro-optic and radio frequency countermeasure systems [19, pp. 18–19]. The open architecture allows the aircraft ADPROS to communicate to either other aircraft or ground vehicles relaying threat information and to leverage the other platforms to employ responses to defend the air platform. Figure 1 depicts how the aircraft survivability system works to protect the aircraft and relay threat information to other platforms.

The open architecture connectivity enables the employment of other air and ground systems to create a multi-tiered threat defeat capability. The multi-tiered defeat system of systems integrates the multiple air and ground survivability systems allowing them to collaborate, detect, identify, and defeat threats encountered on the battlefield [19, p. 19]. US Army aircraft need a lightweight computing system that presents a high processing speed and will enable future autonomous decision-making. Only nanotechnology, quantum computing, and AI can enable the technological solution to fulfill the ADPROS need. Figure 2 depicts how the multi-tiered defeat construct passes critical threat information enabling the system to leverage all battlefield systems to defeat threats.

The problem is that, though the current technological innovations can replicate this capability, these systems are too large and heavy to place on aircraft or ground vehicles. The processing speed of current computers cannot run the algorithm to derive a suitable survivability solution in time to protect the air platform from a threat. To achieve this capability the US Army needs to focus S&T investments in the foundational innovations of nanotechnology, quantum computing, and AI to not only achieve this asymmetric capability but to create the foundation for the innovations that the US Army outlined in the ARCIC document, "The Warfighters' Science and Technology Needs." A foundational innovation is a science and technology that enable progress and applications in a variety of domains; in other words, a technology that serves as the building block for other technologies [20].

The three foundational innovations of nanotechnology, quantum computing, and AI are critical to achieving the US Army's requirements, and they may be applied to solve the technological problem. Nanotechnology is the first critical ADPROS capability needed to not only construct smaller sensors and computers but to enable the system to function as Army Aviation needs. Nanotechnology, specifically, nanomanufacturing, is the true foundational innovation necessary to produce the solution the

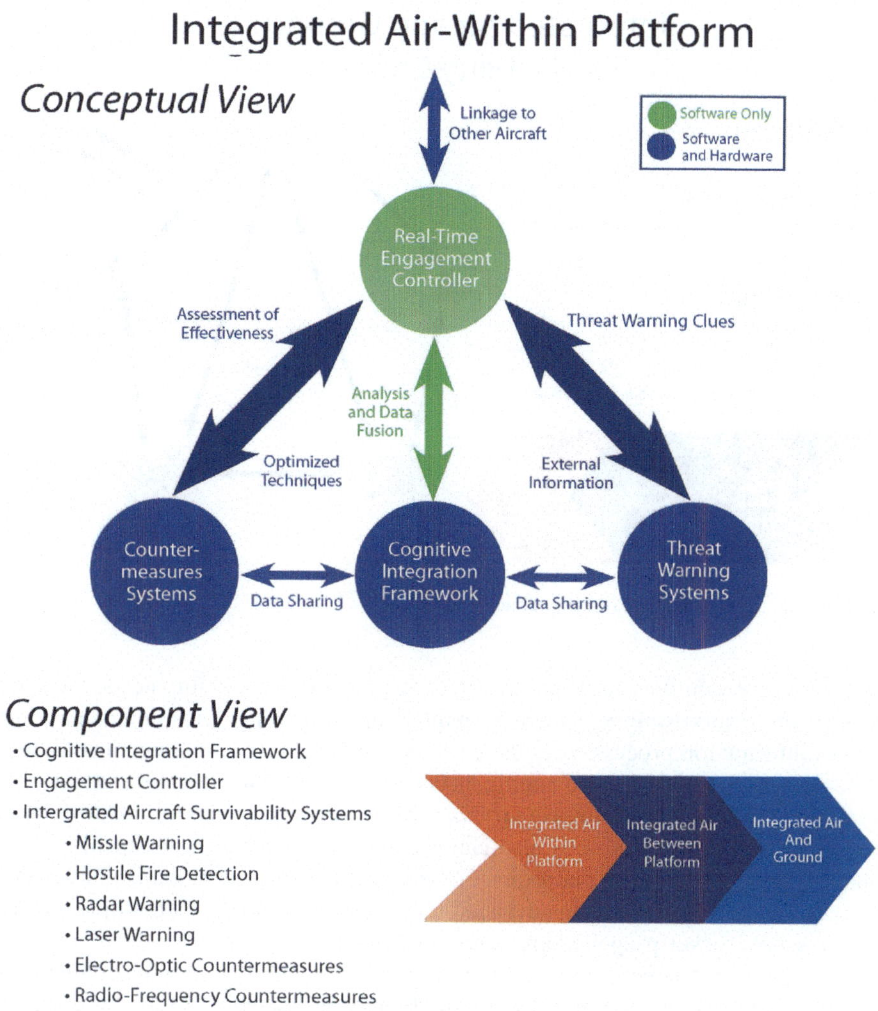

Fig. 1 Onboard aircraft open network survivability system [19]

US Army is seeking and this is where the Army should invest first to achieve the ADPROS capability described above. Manufacturing miniature machines for computer technology is the initial step to enabling supercomputers and AI to operate an integrated survivability system, but the process to self-assemble at the nano-level is where the current scientific struggle is occurring [21]. Solving this problem through investments opens the door to the other technologies the Army needs to increase its technological capacity. Nanotechnology is the building block for the future, but specifically, it is necessary to create a small, but powerful quantum computer [22].

Army Aviation needs a computer capable of rapidly executing the algorithms enabling a network of individual air/ground systems that share information on an

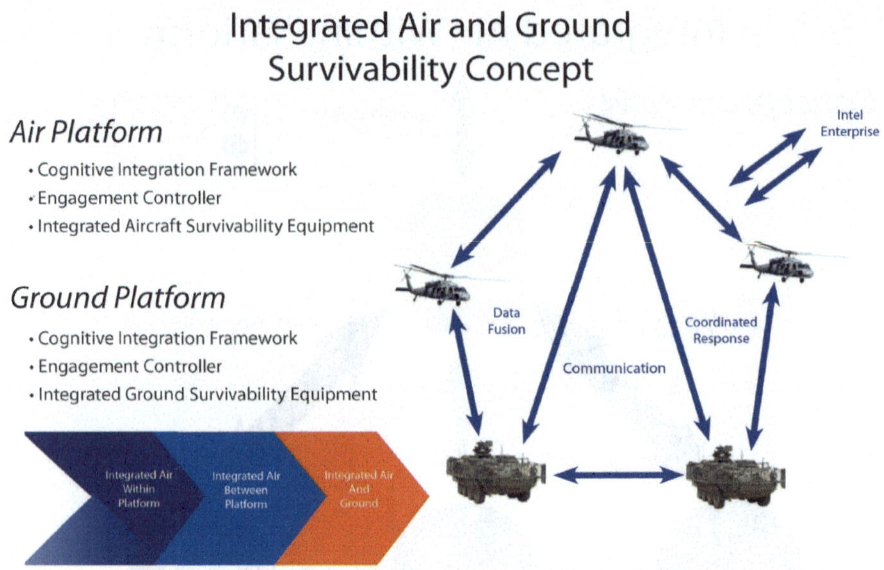

Fig. 2 Open architecture survivability concept [19]

integrated survivability system [19, p. 19]. The processing capability needed to synchronize these survivability systems is in quantum computing. Quantum computing enables information processing at the scale beyond current computing capabilities, while AI is the autonomous and intelligent capability needed to integrate and analyze, develop solutions, and determine a survivability action to protect the air platform [19, p. 17]. Further, nanotechnology also can produce two other critical capabilities to enhance Army Aviation air platforms. For the ADPROS to work, it must be able to detect the threat and requires an advanced sensor array able to detect multiple threats from missiles, rocket-propelled grenades, and bullets.

Recently there has been a breakthrough in sensor capability with the application of nanotechnology melded with Graphene. Graphene Flagship, a European Union organization, has taken a significant step in thermal infrared (IR) photodetectors with the development of the most sensitive uncooled graphene-based thermal detector yet fabricated [23]. These new photodetectors, known as bolometers, are so sensitive that they can register the presence of a scant few nanowatts of radiation, which is the equivalent of a thousandth of waving a hand in front of the detector [24]. Seeing is not only critical for the ADPROS to operate appropriately, but also if all the systems fail to stop the threat, nanotechnology can improve survivability through the development of lightweight structural components and skin that have an anti-ballistic capability [25]. The key is to harness the breakthroughs within these foundational innovations enabling the creation of disruptive innovation that ensures a strategic asymmetric advantage and continues US Army global military dominance.

To analyze how the foundational innovations of nanotechnology, quantum computing, and AI impact the US Army's future dominance we must have a better understanding of the impacts of different types of innovation. Disruptive innovation has been described as a process by which a product or service takes root initially in simple applications at the bottom of a market and then relentlessly moves up market, eventually displacing established competitors [26]. A prime example is the use of personal computers. Initially, computing technology was expensive and complicated and through the application of technological advancements in computers made them cheaper and better, which translated to increased accessibility to consumers. Another example is the cell phone, which disrupted the fixed-wire phone industry, creating a market that disrupted and altered how we live today. The industry that used innovation to improve fixed-line telephone capability could not adapt to the disruption caused by cheaper cell service.

Disruptive innovation highlights competition between organizations, which also applies to militaries. Regarding the application of military power, the business model needs further refinement for military application. All militaries have objectives and metrics for success, which drive innovation to improve their capabilities. Generalized Disruptive Innovation (GDI) theory is an adaptation of disruptive innovation for military applications which takes into account that military competition involves violence as a metric of success [27, p. 128]. GDI is a theory of organizational competitioin which is appliec to innovations that improve competitive organizations' performance in tasks they have historically considered of lesser importance but not tasks of primary importance [27, p. 131]. Such a model applies directly to the current US Army's efforts, as it invests in research in new areas such as advanced protection capabilities, networking, Robotic and Autonomous Systems (RAS), Soldier/Team Performance, Expeditionary Mission Command, and Cyber Electromagnetic. The US Army intends to create a new asymmetric advantage but not to diminish its requirements for sustaining military dominance as a warfighting force [10, pp. 9–13]. Competition forces state and on-state organizations to improve military capability by making sustaining innovations to match near-peer competitors.

The sustainment innovations in which the US Army is investing are the FVL and Combat Vehicle Modernization Strategy to maintain the current dominating military capacity required to achieve its primary task to win the nation's wars. When the US Army demonstrated the might of its military power in the Persian Gulf War, this was a display of military GDI. The application of extended-range precision-guided munitions, stealth aircraft, advanced rotary wing aircraft, and ground vehicles was a disruptive innovation globally. Nations such as Russia and China began investing in innovation to match and counter the new military innovation that changed the paradigm for the application of military technology and dominance.

Since 1991, the US Army has engaged in sustaining innovations to maintain military dominance with improvements to the current legacy platforms, but, as detailed previously, the asymmetric advantage is shrinking due to the investment efforts of competing state and non-state actors. As the US Army is re-evaluating its current capabilities against the changing global landscape, it has laid the foundation for the leap in asymmetric military dominance as it enters another RMA. The United States is investing in the next generation of technology to achieve the 3rd offset in military capability [3]. The US Army is searching for the disruptive innovation opportunities within its investment strategy to maintain the strategic asymmetric advantage and ensure US military freedom of movement and global dominance.

So why is GDI essential to strategic asymmetric advantage? Investments within a GDI framework yield the advancements in technology required to reproduce the asymmetric advantage found on the battlefield during the Persian Gulf War. The strategic asymmetric advantage is acting, organizing, and thinking differently than opponents to maximize one's advantages, exploit an opponent's weaknesses, attain the initiative, or gain greater freedom of action [28]. Applying the GDI theory allows the US Army to create a new market in warfare while this asymmetric advantage forces state and non-state actors to re-evaluate their military investment strategy to re-tool and match the new emerging disruptive capability the US Army is currently investing to achieve.

5 Evaluating the Disruptive Innovations

The next industrial revolution purportedly will yield leapfrog technology in AI, robotics, the Internet of Things (IoT), autonomous vehicles, 3-D printing, nanotechnology, biotechnology, materials science, energy storage, and quantum computing [16]. So why are nanotechnology, quantum computing, and AI considered a GDI? The answer is that these three technological innovations are the foundation for the above listed emerging technologies and have the capability to produce the innovation needed to achieve the Army Aviation's ADPROS material requirements.

As stated before, nanotechnology serves as the foundational capability and overcomes the size/weight restriction. It enables the quantum computing capability necessary to process the algorithms and communicate via a network with other ADPROS systems. It facilitates an AI autonomous decision to execute an ADPROS solution for Army air platforms. Additionally, nanotechnology can increase air platform survivability through the application of ballistic protective nanomaterial. Nanotechnology combined with a quantum computer and AI is the gateway to creating the next GDI, yielding continued US Army dominance. These capabilities applied to a US Army air system mitigate shoot-down risks, which allow US Army ground forces global freedom of maneuver, creates a strategic asymmetric advantage. This study supports this by applying a GDI-based definition of these foundational technologies to evaluate the potential for disruptive innovation and strategic asymmetric advantage. The competitive organizations are the US Army and other state and non-state competing militaries.

The Army is applying sustaining technologies to maintain the current military dominance, but the increasing global threat and capabilities of state and non-state military organizations is eroding the US Army's dominance gap. The US Army's main task is to fight and win its nation's wars. To counter the capability erosion, the US Army has set forth on the route of historically lesser importance by focusing on Soldier/Leader development, a new Army Operating Concept, and through investment in emerging technology. The less significant innovative path the US Army

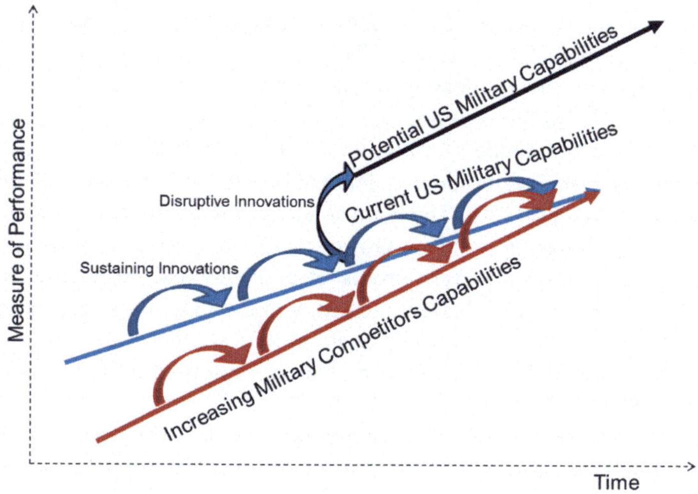

Fig. 3 Application of GDI to current US Army capabilities [36]

is pursing is through emerging technologies. As with all disruptive innovations, these technologies do not yield an immediate military capability but are the source of future required systems capabilities needed to produce the strategic asymmetric advantage. Figure 3 depicts the impacts of investing in the foundational technologies of nanotechnology, quantum computing, and AI graphically. This disruption creates a new path of military capability that is beyond the capabilities of other military competitors.

The US Army's investment strategy aims to improve existing military capabilities through sustaining innovations as well as to make the next leap by leveraging a disruptive innovation. Nanotechnology, quantum computing, and AI are the leading options to leverage the disruptive innovation for Army Aviation, but they will have a far-reaching influence in all US Army military capability outcomes. Investment in these three technologies supports the innovation required for future Army Aviation ADPROS, and it will affect every future Army system needed to maintain the US Army's global military dominance.

6 Recommendations

> The best way to predict your future is to create it. Abraham Lincoln [29]

The author recommends that the US Army invest within nanotechnology, specifically in nano self-assembly and in nano-manufacturing, as these areas are currently undergoing impediments to progress and could use investment. As noted by George Tulevski, the issues with nanotechnology are building components for nano-machines at the particle level. Because of this problem, advancements in nanotechnology development have slowed. Tulevki's solution is to have the particles self-assemble through the application of chemistry. Chemistry allows the particles to self-assemble to build the necessary structure to enable the technology [21].

The other area recommended for investment, nano-manufacturing is also experiencing a pending gap in development as detailed in a 2014 US Government Accountability Office (GAO) report [30]. The main concern of the industry experts consulted by the GAO is "the missing middle." Government, universities, and startups focus their investment on basic research, proofs of concepts and production at laboratory scale. However, the GAO believes that to win the global nanotechnology race, America must bridge the gap between such activities and be able to produce the technologies at scale, including everything from prototyping to developing reliable manufacturing processes [31].

Focusing US Army S&T nanotechnology investments to not only solve the self-assembly problem but also concentrate on developing the reliable manufacturing process sets the foundation to create the breakthroughs needed and to shape the development cycles to field an Army Aviation ADPROS before 2030. Based on the current development cycles within nanotechnology, the National Aeronautics and Space Administration (NASA) expects that between 2019 and 2027 the computing technology, sensors, and airframe structures needed to create the US Army's S&T needs for ADPROS will be available [32].

Breakthroughs in nanotechnology are the foundational innovation for quantum computing and AI, and these two innovations are critical to the disruptive innovation necessary to change the US Army's current military capability. Additionally, the author also recommends investing in quantum computing and AI. All three are critical for robotics, IoT, autonomous vehicles, 3-D printing, biotechnology, materials science, and energy storage necessary to fulfill the US Army's identified Warfighters' S&T needs [10, p. 3]. Nanotechnology development yields the quantum computing capability required in producing the necessary capacity, and together they enable the Army's S&T needs for ADPROS around 2025 [33].

Besides the hardware development for quantum computers, the investments need to focus on algorithm development. This quantum function is critical to enable AI. AI needs the capabilities generated by nanotechnology and quantum computing. Current AI capability we see in everyday life is known as Narrow AI. Narrow AI enables language translation, self-driving vehicles, and image recognition [34, p. 7]. As Narrow AI continues to increase in capability, it is a critical component to Army Aviation's ADPROS, but to achieve the full potential of a disruptive innovation Army Aviation needs Artificial General Intelligence (AGI). AGI refers to a notional future AI system that exhibits apparently intelligent behavior at least as advanced as a person across the full range of cognitive tasks [34, p. 7]. Since this capability will not be available until beyond 2030, the US Army should focus investment in modern machine learning under Narrow AI. Modern machine learning is a statistical process that starts with a body of data and tries to derive a rule or procedure that explains the data or can predict future data [34, p. 8]. Modern machine learning is the final piece of the puzzle to bring ADPROS to its full potential by creating an airframe protection system enabling the US Army global freedom of maneuver.

Other areas recommended for further research or action that play a critical role in maintaining US Army military dominance include reforms to the Army Acquisition process to accommodate and incorporate technological advances in developing systems to keep pace with the rapid rate of projected technological development. Additionally, the author recommends the US Army track innovations and technologies that industry may derive from any US Army investments within nanotechnology, quantum computing, and AI technologies to protect taxpayer investments and government use rights. For example, the US Army spends funds for the development of technology after which the Army obtains government use rights over the technology.[1] Having these rights is critical to ensure the US Army maintains access to critical engineering data, which allows the US Army to mitigate future costs by having the ability to conduct rebuild, reconstruct, and execute repairs within Army Depots or through US military labor.

7 Conclusions

> Without change, there is no innovation, creativity, or incentive for improvement. Those who initiate change will have a better opportunity to manage the change that is inevitable. William Pollard [35]

The US Army is transitioning its force to maintain the asymmetric advantage it achieved and displayed during the 1991 Persian Gulf War. The overwhelming display of military power spurred state and non-state actors to undertake sustainment innovations to match the US Army's dominant capability. The US Army is at a critical crossroad and must take decisive action to maintain its asymmetric advantage. It needs to rapidly leverage emerging technologies balanced with soldier development and adaptive doctrine to dominate in a fluid future OE. A new disruptive innovation will sustain US military dominance beyond current technologies. The US Army can achieve the necessary General Disruptive Innovation by investing in critical foundational technologies catapulting the US Army into an unmatched asymmetric strategic advantage in military capability equal to the capacity displayed in 1991. These foundational innovations are nanomanufacturing, quantum computing hardware and algorithm development, and advancements in narrow AI modern machine learning. These three technologies are not only the building blocks for Army Aviation ADPROS needs but will have a profound effect on all identified US Army future capability requirements. Focusing investments within these foundational innovational areas, the US Army will produce greater outcome at a reduced expense, creating a GDI and a strategic advantage for the United States, resulting in global military freedom of maneuver and continued US Army dominance as the most powerful ground force in the world.

---

[1] Rafael Rosado, Attorney Advisor, AMC Legal Center, Army Material Command, email interview by author, January 30, 2017.

References

1. Sinquefield R (2013) Innovation is key to the success of 21st century philanthropy. Forbes, 16 May 2013. http://www.forbes.com/sites/rexsinquefield/2013/05/16/innovation-is-key-to-the-success-of-21st-century-philanthropy/#75f73a5a29f0
2. Schubert FN, Kraus TL (eds) The whirlwind war. Center of Military History United States Army, Washington DC, p 25
3. Hagel CT (2014) Reagan national defense forum keynote. Public speech, Ronald Reagan Presidential Library, Simi Valley, CA, 15 Nov 2014, p 6
4. Lissner RF (2016) The long shadow of the gulf war, 24 Feb 2016. https://warontherocks.com/2016/02/the-long-shadow-of-the-gulf-war/
5. Spada A (2016) Airplanes at risk: Daesh has new military technologies? 7 Jan 2016. http://www.islamedianalysis.info/airplanes-at-risk-daesh-has-new-military-technologies/
6. Horowitz MC (2010) Military power and major military innovations. In: The diffusion of military power, causes and consequences for international politics. Princeton University Press, Princeton, p 22
7. Stephenson S (2010) The revolution in military affairs: 12 observations on an out-of-fashion idea. Mil Rev 38
8. U.S. Department of the Army (2014) The U.S. Army operating concept: win in a complex world, TRADOC pamphlet 525-3-1. U.S. Army Training and Doctrine Command, Fort Eustis, VA, 31 Oct 2014
9. Howard M, Guilmartin JF (1994) Two historians in technology and war, 1 July 1994. http://www.strategicstudiesinstitute.army.mil/pubs/display.cfm?pubID=354
10. U.S. Department of the Army (2016) The warfighters' science and technology needs, 21 Sept 2016
11. Bender J (2016) The US intelligence chief made some chilling predictions for 2016's biggest global threats, 10 Feb 2016. http://www.businessinsider.com/these-are-the-main-global-threats-for-2016-2016-2
12. U.S. Department of the Army (2016) Future force development strategy, 23 Nov 2016 (Pre-decisional Draft)
13. Youssef NA, Harris S (2016) U.S. fears new missile race in Syria would help ISIS, 25 Apr 2016. http://www.thedailybeast.com/articles/2016/04/25/u-s-fears-new-missile-race-in-syria-would-help-isis.html
14. Lindell J (2009) Clausewitz: war, peace and politics, 26 Nov 2009. http://www.e-ir.info/2009/11/26/clausewitz-war-peace-and-politics/
15. U.S. Department of the Army (2012) Mission command, ADRP 6.0. U.S. Army Training and Doctrine Command, Fort Eustis, VA, 17 May 2012
16. Schwab K (2015) The fourth industrial revolution, what it means and how to respond. Foreign Aff. https://www.foreignaffairs.com/articles/2015-12-12/fourth-industrial-revolution
17. Army Warfighting Challenges, 12 Oct 2016. http://www.arcic.army.mil/App_Documents/AWFC-Current.pdf
18. Trent Lythgoe, US Army (2015) Advanced protection systems. In: Information paper, ARCIC, 7 Dec 2015
19. Calafut M (2015) The future of aircraft survivability. Army Technol 3(2)
20. Chameau JL, Ballhaus WF, Lin HS (eds) (2014) Emerging and readily available technologies and national security: a framework for addressing ethical, legal, and societal issues. National Academies Press (US), Washington (DC), p 31
21. Tulevski G (2016) The next step in nanotechnology. Ted Talks, video file. https://www.ted.com/talks/george_tulevski_the_next_step_in_nanotechnology
22. Hardesty L (2016) Toward practical quantum computers, 8 Aug 2016. http://news.mit.edu/2016/toward-practical-quantum-computers-0808
23. Lloyd S (2017) Ultrahigh sensitivity graphene infrared detectors for imaging and spectroscopy. Graphene Flagship, 31 Jan 2017. http://graphene-flagship.eu/graphene-pyroelectric-bolometers-(2)

24. Johnson D (2017) Graphene infrared eye needs no signal amplification, 3 Feb 2017. http://spectrum.ieee.org/nanoclast/semiconductors/devices/graphene-infrared-eye-needs-no-cooling
25. Andrei M (2012) Nanomaterials to prevent speeding bullets, 8 Nov 2012. http://www.zmescience.com/science/nanotechnology-science/nanomaterial-armor-mit-08112012/
26. Christensen C (2012) Disruptive innovation explained, 30 Mar 2012. YouTube, video file. https://www.youtube.com/watch?v=qDrMAzCHFUU&t=324s
27. Mukunda G (2010) We cannot go on: disruptive innovation and the first world war royal navy. Secur Stud 19(1):128
28. Metz S, Johnson DV (2001) Asymmetry and US military strategy: definition, background, and strategic concepts, Jan 2001. http://strategicstudiesinstitute.army.mil/pdffiles/PUB223.pdf
29. Good Reads Abraham Lincoln Quote. The Best Way to Predict Your Future is to create it. http://www.goodreads.com/quotes/328848-the-best-way-to-predict-your-future-is-to-create
30. 2014 Annual Report. Additional opportunities to reduce fragmentation, overlap, and duplication and achieve other financial benefits, GAO-14-343SP, 8 Apr 2014. https://www.gao.gov/products/GAO-14-343SP
31. Nanomanufacturing in America, small but imperfectly formed, 7 Feb 2014. http://www.economist.com/blogs/babbage/2014/02/nanomanufacturing-america?fsrc=scn/gp/wl/bl/smallbutimperfectlyformed
32. 2015 NASA Technology Roadmaps, TA 10: Nanotechnology, pp 42–84, 99–116, July 2015. https://www.nasa.gov/sites/default/files/atoms/files/2015_nasa_technology_roadmaps_ta_10_nanotechnology_final.pdf. Accessed 10 Feb 2017
33. A leap forward for quantum computers, 23 Feb 2016. http://www.idquantique.com/quantum-computers/
34. Preparing for the future of artificial intelligence, 12 Oct 2016. https://permanent.access.gpo.gov/gpo75059/preparing_for_the_future_of_ai.pdf
35. Pollard W (2017) BrainyQuote.com, Xplore Inc. https://www.brainyquote.com/quotes/quotes/w/williampol163245.html
36. Christensen C Disruptive innovation. http://www.claytonchristensen.com/key-concepts/

Revisiting the Revolution in Military Logistics: Technological Enablers Twenty Years on

Zoe Stanley-Lockman

Abstract This chapter builds on the Revolution in Military Logistics (RML) concept, as defined by the U.S. Army in 1999, by conceiving of logistics transformation as a prerequisite for any Revolution in Military Affairs or broader force transformation at the tactical, operational, and strategic levels. Looking past the tip of the spear, this chapter analyses how emerging critical technologies may facilitate a paradigmatic shift in land-based combat service support emphasizing greater agility, modularity, seamlessness, control, and sustainability. The chapter considers technological enablers—including artificial intelligence, alternative energy technologies, augmented reality, additive manufacturing, robotics, blockchain and the Internet of Things—which may make logistics more efficient and effective, yet which have only been developed or fielded in an ad hoc way. With a view toward integrating the technologies into operational concepts and organisational constructs, the chapter assesses the merits, as well as vulnerabilities and cultural difficulties, of transforming future military logistics.

1 Introduction

Twenty years ago, then-Army Chief of Staff General Dennis Reimer announced a bold proposition for force transformation: "there can be no revolution in military affairs (RMA) without first having a revolution in military logistics (RML)." [1]. The conception of logistics transformation as prerequisite to overall force transformation is consistent with literature that emphasises the connections between logistics and strategy. Where strategy is an "art of the possible," [2] logistics is the complementary science of planning, implementing, maintaining, sustaining, and concluding actions. The connections between logistical undertakings and strategy form the vital, if not forgotten, backbone to operational effectiveness. As stated by Mark Erbel and Christopher Kinsey, logistics "determines the variety of strategic options available,

Z. Stanley-Lockman (✉)
S. Rajaratnam School of International Studies, Singapore, Singapore
e-mail: iszlockman@ntu.edu.sg

M. E. Kosal (ed.), *Disruptive and Game Changing Technologies in Modern Warfare*, Advanced Sciences and Technologies for Security Applications,
https://doi.org/10.1007/978-3-030-28342-1_11

but not which strategic course of action will be followed or how effectively it will be implemented." [3].

Furthermore, combat support elements are necessary across the entire gamut of military tasks from low-end enforcement to high-intensity conflict and from force projection to sustainment of operations [4]. Achieving the RML to which General Reimer referred means integrating three domains: "technology application and acquisition agility, force projection, and force sustainment." [5]. Today, integration of these domains means conceptualising the entire lifecycle—from acquisition and development, to training, to deployment and sustainment, and troop drawdown and equipment disposal—as an interconnected process. However, an RML as such has been overlooked for the better part of two decades. This chapter focuses primarily on the first RML domain, technology applications, to assess how emerging technologies can augment the effectiveness of military logistics today.

As detailed below, the RML is composed of six tenets that, combined, can help armed forces achieve an operational or strategic advantage. Each relies, to a degree, on using information technology (IT) systems to have more accurate information delivered in a timely manner across the entire supply chain. The introduction of IT systems has led to incremental innovation for military logisticians, such as the 2003 "Sense and Respond Logistics" initiative, which sought to transform combat support by retooling organisational structures and information systems to better support the network-centric warfare concept [6, 7]. Today other emerging technologies could breed game-changing military logistics transformation in line with changing strategic priorities, modular combat support concepts, and evolving manpower requirements. The idea of leveraging emerging technologies to achieve an advantage is consistent with recent US defence innovation efforts as embodied in the third offset strategy of the Obama administration and the 2018 National Defense Strategy [8, 9], yet the logistical dimension is more often a footnote than the focal point.

Logistical applications of technological innovations have only been considered in an ad hoc manner. Military logistics transformation was well covered in line with the shift from supply- to distribution-based logistics (DBL) in the late 1990s and from platform- to network-centric warfare in the mid-2000s [10–12]. Even in the most recent logistics roadmap, the Defense Logistics Agency (DLA) Strategic Plan 2018–2026, the only technologies mentioned are commercial IT and cloud-based solutions, as well as predictive big data management systems [13]. While important, the focus on information management systems excludes logistics transformation that other emerging technologies may enable. This is consistent with a gap in military logistics literature about the potential of emerging technologies—including artificial intelligence (AI) and machine learning, robotics, additive manufacturing, blockchain, alternative energy technologies, and the Internet of Things (IoT)—to revolutionise logistics today.

Military logistics is also ripe for transformation due to a long history of privatisation and cooperation with the private sector, processes which have been mimicked across the whole of military organisations today to varying degrees of success. The longstanding liaisons between military logistics and contractors have been accelerated as combat support elements seek to decrease stockpiles of "just-in-case" supplies toward leaner delivery of "just-in-time" logistics interventions. Supply-based stockpile management was attractive to conscript militaries because it typically required less training. Having a professional military also meant more specialisation could be possible—and was indeed necessary given the increasingly sophisticated military equipment that all military departments acquired. As such demand-oriented supply chain management was ripe for 'professionalised' supply chain management, including roles for logisticians with more technical skillsets.

With professional militaries in mind, this work argues that emerging technologies will increasingly find military logistics applications. Downsizing of logistical forces and organisational shifts toward smaller, more modular units also make emerging technologies attractive. However, even if they are used to compensate for decreasing manpower, adoption of emerging technologies will necessitate changes in training, organisational structures, and command and control (C2) for logisticians.

2 Thought and Context Behind a Revolution in Military Logistics (RML) in the 1990s

The backdrop to initial discussions of a Revolution in Military Logistics (RML) dates back to 1990–1991 with Operations *Desert Shield* and *Desert Storm*, the largest-ever rapid deployment of U.S troops and supplies at that time. Hereafter referred to as the First Gulf War, these operational experiences are recognised as the culmination of the Revolution in Military Affairs (RMA) from the 1970s and 1980s [14]. The First Gulf War also forms an important backdrop to the discussion because, despite several strategies and initiatives, shifts to logistics since 1991 have been more incremental rather than paradigmatic [7, 15].

2.1 Lessons from the First Gulf War

Although considered an overwhelming operational success in the short term, the First Gulf War also manifested the logistical difficulties that combat service support (CSS) would have to surmount in the immediate aftermath of the Cold War. As recounted in a Government Accountability Office (GAO) report on supply efforts for land-based forces during the First Gulf War, the logistical successes resulted from workarounds rather than from an execution of the doctrine at the time [16]. For instance, the unusually high aircraft readiness rates of 90–97% were attributed to: supplying parts (particularly in the form of War Readiness Spares Kits) to units, trading parts with "sister battalions" or other collocated units, and reusing parts from non-mission capable equipment [16]. Dependence on the local supply chain, including for delivery of logistics on the ground, became a necessary workaround without which logistical shortcomings would have been more severe—even for a short missions. The fact that these supplying practices supplanted, in most cases, direct use of the Department of Defense (DoD) supply system based in the continental US illustrates success of battlefield decision and in-theatre innovativeness, rather than a predictable, replicable, and systematic approach to logistics.

Many of these shortcomings can be attributed to a supply-based logistics chain that overemphasised just-in-case stockpiles. Heavy lift for vehicles was deemed insufficient, leading to other logistical shortcomings such as inadequate reinforcements and dependence on local forces for "fuel, water, food, shelter and communications." [17]. While information about the availability of parts could be ascertained within 24 hours [16],[1] transportation of those parts could take weeks—at which point they were often delivered to areas where mobile maintenance personnel were no longer located. Furthermore, US Marines reported difficulty managing stockpiles of unused supplies that had been prepositioned. A majority of the requisitions had never been requested: for instance one official noted that only 800 of the 1,800 line items the US Marine Corps (USMC) delivered by cargo ships were relevant to the mission [18].

A further issue was asset visibility and control. The USMC had difficulty computing capability shortfalls to track equipment. Because Central Command sent and deployed combat troops before support personnel arrived, personnel who were needed to manage equipment arrivals and inventories were otherwise preoccupied in-theatre, resulting in lost items and insufficient visibility and control. Overall, missing or incomplete shipping records meant that supply officers only had a "general idea" of what was being delivered [19].

Flexibility in sourcing and utilising in-theatre assets, as demonstrated by forces in the First Gulf War, will always be required, as operational contingencies are an unescapable reality. Nonetheless, logistical shortcomings during these operations forebode a new reality: the logistics paradigm from the Cold War would need to be revolutionised. If these were the types of issues personnel would encounter in short, successful missions, then the risk of these supply-based logistics shortcomings being further amplified in protracted missions with superior sustainment requirements was too great.

The Army logistics community hence commenced ambitious initiatives to revolutionise logistics. Critical among these was the "Focused logistics" model, the logistics component of the Force XXI Advanced Warfighting Experiment (AWE) from 1997. Then-Lieutenant Colonel Douglas MacGregor described Force XXI as a way for the Army to respond to two characteristics of future warfare.

First i was keeping information dominance as the foundation of future warfighting [20], which included digitizing Army divisions. In this vein, Force XXI logistics sought to redefine the logistics paradigm by "[harnessing] information technology and related management principles towards the goals of increasing responsiveness while reducing the cost and "footprint" of logistics forces in a theater of operations."

[1]The information could be sufficiently processed in a 24-hour span as of January 1991, when Air Force personnel operationalised a new computing system that centralized data and provided visibility on spare parts. See [16].

[21]. Thus began the shift from supply- to distribution-based logistics, which would eventually become one of the RML tenets carried through to Operation Iraqi Freedom.

Similar to the shift from supply-based logistics where everything would be available "just in case", the move toward a DBL model placed the emphasis on demand by depending on better data about what types of requisitions would need to be delivered "just in time".

Second i was new military technologies expanding battlespaces in such a way that would require doctrine and strategy to be adaptive. Already in the Gulf War, DoD encountered issues relating to long supply lines. MacGregor's point that new weaponry "with increased range, accuracy and lethality" [20] will impact doctrine and strategy by altering the very definition of "frontlines" also implies a logistical challenge: in an expanded battlefield, "last mile" supply, often the most difficult phase of logistics delivery, is commensurately enlarged. Furthermore, urban warfare and "wars among the people" which would become increasingly prevalent in the succeeding years also complicate logistics systems. The "absence of clearly identifiable safe areas, lines of communications, and front lines" in ill-defined battlespaces, as argued by Erbel and Kinsey with regards to post-2003 operations in Iraq and Afghanistan, would come to exemplify the difficulty of traditional logistics supply missions [3].

Beyond battlefields, battlespaces also include attacks in cyberspace or space, as well as the militarisation of traditionally civilian spaces. While important, the logistical tails as related to these domains are beyond the scope of this chapter.

2.2 *A Special Issue Calling for More Specialisation*

With recognition of the First Gulf War logistical shortcomings and foresight about future warfare in mind, the U.S. Army published a special issue of its magazine, *Army Logistician*, titled "The Revolution in Military Logistics" in 1999. The special issue does not represent the first invocation of the term "revolution in military logistics." However, it is the clearest articulation of the concept. As such it forms the basis for the remainder of this chapter.

The six tenets of an RML—"a Seamless Logistics System, Distribution-Based Logistics, Agile Infrastructure, Total Asset Viability, Rapid Force Projection, and an Adequate Logistics Footprint" [5]—mostly remain conceptually relevant today. The first is a *seamless logistics system*, benefitting from interconnected and interoperable IT systems and networks. *Distribution-based logistics*, perhaps the most visible in military thought in the years to come, is characterised by the shift away from stockpiles toward end-to-end distribution of logistics as required per each unique mission. Although this would later be altered, DBL had been referred to as the "envisioned RML end-state." [22] "Agile infrastructure," interchangeable with *agility*, emphasises a modular approach to the scale, size, technical expertise, teaming, training, and manoeuvre of logistics. Next is *total asset visibility and control*, which involves tracking the status and location of the supply chain and unit requirements in real-time. The fifth pillar, *rapid force projection*, involves the capabilities needed to project strike and sustained manoeuvre forces, as well as intra-theatre mobility. Lastly, as addressed in the conclusion, RML requires an *adequate logistics footprint*, or the determination to have sufficient CSS for given operations.

However, as noted by Victor Maccagnan, AWE and initial RML initiatives merely resulted in a few think pieces and failed to generate more noteworthy analytical debate or strategic changes [7]. While these few articles, including the *Army Logistician* special issue, were prescient in many ways, the attacks of 11 September 2001 severed any remaining bandwidth to effectuate a RML with more pressing priorities at hand. By 2005, Maccagnan cited logistics as the area least affected by revolutionary force transformation [7].

Subsequent combat experiences in Iraq and Afghanistan show a shift, albeit an incomplete one, away from supply-based, just-in-case logistics. Operation *Iraqi Freedom*, the first battlefield test for DBL, demonstrated the insufficiency of the model in expeditionary environments and over a protracted period. In Operation *Iraqi Freedom*, the DoD lacked adequate CSS personnel and transportation assets, had insufficient visibility on the movement of assets in real-time, suffered interoperability pitfalls, and mismanaged communication between DLA and the boots on the ground [23, 24]. Already by the time the DBL model was deployed, its replacement was being conceived.

Placing more emphasis on sustainment, the subsequent model performance-based logistics (PBL) gained more popularity that persists to this day. Whereas DBL represented the first shift away from supply-based, just-in-case logistics toward demand-based, just-in-time logistics, the main addition of the PBL framework that was missing in DBL is a long-term focus on the entire support lifecycle [25]. Instead of paying part-by-part, PBL aims to reduce logistics and maintenance costs by building those into the front end of the acquisition management cycle.

The core objective of PBL is "to improve system equipment readiness and reduce total ownership costs." The overall aim of DBL, as stated by David Payne, is to "enhance responsiveness of support, and reduce the logistics footprint simultaneously." [26]. The importance of total ownership costs cannot be understated: operations and support costs make up 60–70% of total lifecycle costs, with maintenance making up an increasingly large portion of total logistics costs [27, 28].

Although the two models do not share the same metrics, they do share mutually compatible aims of enhancing CSS effectiveness and lowering logistics costs, as will be revisited later in this chapter. Nonetheless, it is worth noting that, because of the lifecycle focus embedded into PBL, it offers more room to discuss the integration of dual-use and emerging technologies into weapons systems in a way that DBL would have more trouble measuring over a period of time.

While logistics transformation focus is distilled into literature on DBL or PBL, the other tenets of an RML appear to only have advanced and received attention in an ad hoc manner. Revisiting these tenets 20 years later, the concepts are still largely applicable, and several of the technologies then in research and technology (R&T) or research and development (R&D) phases are reaching maturation. Even if a formal, strategic approach to effectuating a RML dropped off some-twenty years ago, this chapter examines the military applications and deployments of such technologies to assess progress toward realising an RML today.

3 Technological Enablers Coming of Age for a Contemporary RML

3.1 Seamless Logistics System

Envisioning the end-to-end integration of "the logistics management framework, command and communications processes and automation architecture," [29] a seamless logistics system acts across all logistics phases, translating operational concepts into the management, delivery, replenishment, and disposal of logistical assets. The integration must be vertical, connecting tactical to strategic levels through connections to command and control (C2), as well as horizontal, across logistics functions through sensors on weapons systems. Technologically, this means robotics and big data are vital to realising this architecture, as are information and communication technologies (ICT) also described below.

Once operational decisions are made at the strategic levels, a seamless logistics system should be able to automatically translate the decisions into logistical tasks. Instead of calculating which, how many, and when assets should be deployed in-theatre, computerised models could show how certain logistics packages constrain or enable action. If effective, the instantaneity of computer modelling and simulations could make logistics a stronger decision-making factor, rather than a consequence of capability planning decisions, in C2 systems. While developing military campaigns is an inherently iterative process, C2 systems often dictate logistical tasks for operations in a top-down fashion. In the way that domain awareness is vital to planning and executing military tasks, logistical awareness is a key determinant in their success. Real-time modelling could ensure greater synergies between services and the combatant commands for in-theatre logistics.

Machine learning could help logistical models evolve, taking into account data from previous missions as the iterative C2 process perpetuates in expeditionary operations. Be it through complex, multivariate regressions or AI algorithms, big data can marry sensors in inventoried and deployed equipment directly with the strategic mission level.

Below the threshold of AI systems, new software management tools are already helping translate data to better informational awareness on availability of assets. In Qatar, the U.S. Combined Air Operations Center recently began using a software management tool, known as JIGSAW, which streamlines data from signals, communications, and human intelligence sources. Programmed to "[consider] which aircraft can be refueled by which tankers, [give] takeoff and landing times, and [deconflict] airspace," digitising this previously analogue process "saves three to four hours a day" [30] and "[generates] savings of 400,000 to 500,000 lb of fuel—or $750,000 to $1 million—each week." [31]. Anecdotal today, work with the commercial tech sector for solutions like JIGSAW are progressing into a mainstay. The time savings in particular are noteworthy: being able to calculate these specificities with the click of a button rather than on whiteboards, logistics planning no longer has to be a subsequent step to capability planning and other C2 decisions.

Instead, modelling different logistical scenarios can be built into the front end of capability planning for services. When designing force packages, considering the logistical tail at the click of the button makes for more effective operations. In the future, AI models could teach themselves to account for unexpected replenishment needs or redirected assets at select locations or timepoints—therein bolstering sustainment capabilities with more ease in earlier stages of campaign design. Deploying such AI applications for military logistics is a new frontier and can be expected to become more prevalent with the rise of autonomous weapons systems where logistical concerns may outweigh manning ones.

More accurate, timely information about the state of military equipment and networks can also help avoid logistics stovepipes. At the tactical level, horizontal integration between weapons systems can help inform decisions about force package design. The basic foundation for horizontal integration is, and will remain, sensors. Without sensors, which are notably becoming less expensive and more scalable, the data could not be collected to then be exploited. Using sensors, two emerging technology areas that can be expected to create more comprehensive images of equipment's readiness at a given time are Internet of Things IoT) devices and digital twinning.

Data collected from IoT sensors can be used, for instance, to identify causes of excessive fuel burn [32]. With a recent GAO report recommending DoD report more accurate fuel consumption data—which cost the department $10 billion per year [33]—to Congress, the installation of IoT sensors to "inexpensively and precisely monitor tank levels, temperatures, and flow rates" [34] could make fuel management systems more accurate. Other potential logistical applications of IoT sensors, as suggested by the DoD Chief Information Officer in 2016, include: systems integration for more complete information on inventory, usage, and payments; installing motion sensors for enhanced perimeter security; continuous measurements for leaks in platforms and facilities; and remotely operating parts and components from smartphones [34].

In addition to AI algorithms exploiting data collected from IoT devices, as explored below, future facility and platform management systems may also use IoT data to create digital twins, or virtual replicas of physical assets that can be traced and updated throughout the entire acquisition lifecycle [35]. However, while facility management systems aided by IoT devices is attractive, valid concerns about IoT security may render them less attractive for widespread usage. Digital twins may also prove to be prohibitively expensive, particularly given insatiable processing power demands [36]. The risks associated with centralised logistics systems, as expanded upon with regards to F-35 logistics software below, leave little room for error before reparations become exorbitantly expensive.

The exploitation of data collected from sensors is more important than the act of collection itself, meaning horizontal integration ultimately depends upon data management and analytics. One particularly successful data processing method, as described below, is predictive analytics. Like so many digitisation enablers, horizontal integration also amplifies cyber vulnerabilities. Particularly for IoT devices, many of which are designed without sufficient security measures, horizontal integration can mean multiplying the number of entry points for cyberattacks or sabotage. Indeed in a 2018 report, federal agencies named IoT security one of the 26 long-range emerging threats facing the United States, specifically, "…difficulties protecting networks and data as IoT grows and traditional approaches for security (e.g., encryption) may no longer effectively protect information. Adversaries could also disrupt IoT-enabled critical infrastructure and devices." [37].

Cybersecurity and accuracy concerns have already tarnished the reputation of the F-35 Joint Strike Fighter maintenance software, the Autonomic Logistics Information System (ALIS). ALIS, which centralises logistics data from the wealth of sensors aboard the F-35, had issues meeting Initial Operational Test and Evaluation standards due to hundreds of "critical must-fix deficiencies," most of which relate to cybersecurity in recent years [38]. Two years later, issues with the third iteration of ALIS continue to persist and may have led to the compromise of F-35 data. In January 2019, ALIS received an even more scathing review that concluded bad data has led to low trust levels in the system and low mission-capable readiness levels [39]. If the ALIS example is at all instructive, completing the information loop in horizontal integration may prove too costly or even undesirable. The low levels of trust also show that cultural and organisational elements may be even more vital to force-wide integration than the enabling technologies themselves.

Cultural boundaries to new technology adoption should not be understated, although they are not insurmountable. In general, many technologies that impose an initial cultural boundary are already familiar due to their omnipresence in commercial technologies such as smartphones and IoT devices. More concretely, in one case initial scepticism from test engineers and operators over using one 3D-printed flight-critical part was quickly replaced with enthusiasm that led the engineers to identify five other parts ahead of the allotted timescale [40]. In the long run, a self-perpetuating system requiring limited human intervention may be more attractive to operators than initial attempts at digitisation, which still required a heavier hand and greater possibility for human error from having to manually insert information into spreadsheets.

3.2 *Distribution-Based Logistics (DBL)*

Of the six pillars, DBL is the one that has garnered the most attention in the intervening period between RML discussions in the late 1990s and today. DBL is based on the idea that logistics should be tailored to each mission rather than approximated based upon historical demand. As mentioned above, PBL has garnered more attention due to its emphasis on sustainment costs. These models share similar goals: having capabilities that result in higher readiness levels and offering more options to design logistics packages tailored to specific missions. Although the metrics used to evaluate how these aims are achieved differ, both distribution- and performance-based logistics models may rely upon technological enablers to deliver seamless, just-in-time logistics more efficiently. To maintain the focus on the technological dimension and on the RML concept put forth in 1999, this chapter does not compare how the DBL and PBL models price out the emerging technologies mentioned here.

As introduced above, predictive logistics algorithms are particularly promising. The time and cost savings are generated thanks to sensor data—including from IoT sensors [34]—that connects MRO tasks to specific sub-components before they break. Instead of technicians needing to deconstruct equipment based on standardised replacement schedules, AI systems can tell proactively indicate to logisticians and operators when parts will need to be replaced in a more precise, case-by-case fashion. Thus far this is seen with aircraft and land vehicles, with a twofold imperative stemming from cost savings.

Firstly, predictive maintenance is acknowledged as a readiness management force multiplier. By forecasting when exactly logistical support will be needed, the armed forces can benefit from higher equipment readiness rates vital for both force entry and sustainment. A recent Congressional Research Service report denotes a few examples of predictive maintenance measures for both aircraft and land vehicles. In particular, IBM's Watson has been given Army Logistics Support Activity (LOGSA) data to crunch in order to optimise the *Stryker* fleet, and aviation examples whereby SparkCognition's predictive maintenance algorithms show how millions of taxpayer dollars can be saved by pre-emptively identifying when *wear* turns to *tear* aboard Boeing aircraft [41]. SparkCognition algorithms may also offer logistical solutions to other military customers, such as for US Navy Littoral Combat Ships which have experienced turbine issues [42]. These cost savings also allow for more accurate and reliable data that C2 can use to design force packages for specific interventions.

Secondly, higher equipment readiness rates serve to also optimise MRO depots and warehouses. In both public and private sectors, depot management is a core logistical challenge. Predictive maintenance can minimise the amount of time that platforms idle in depots when only specific components or subsystems require MRO. Furthermore, predictive data should notify users when to expect needing the spare parts that would allow the platform to remain operable and the accuracy of algorithms could help reduce the amount of time inspections take. Freeing up resources—both the physical real estate within warehouses as well as the talent that can only maintain specialised systems one-at-a-time—helps streamline logistical processes and continue the shift away from stockpiling or overcrowding warehouses.

In tandem with predictive AI prognostics, robotics could also help complete the shift toward distribution- or performance-based logistics. In addition to augmented reality, discussed in the following section, the breadth of unmanned systems in development and production makes them particularly attractive for distribution tasks because they "can offer tailored approaches rather than one-size-fits-all solutions." [43]. The use of unmanned systems to this effect is not new; however it has become more widespread in recent years.

One notable example is the usage of remotely piloted helicopters in Afghanistan. Following its first successful military tests in 2008, a remote-controlled version of the K-MAX helicopter was deployed in 2011 to airlift up to 6,000 lb (2,722 kilograms) of supplies at a time to remote mountainous regions in Afghanistan. Over the course of two years, the USMC "completed 1,700 day and night resupply sorties, moving around 30,000 lb of supplies each day—a total of around 4.5 million pounds of fuel, food, ammunition and more." [44]. The UK Defence and Security Accelerator similarly funded 25 projects as part of its "autonomous last mile resupply" competition in 2017 [45]. These types of capabilities would not necessarily alleviate manpower concerns, but rather enhance the availability and safety of troops or contractors who would otherwise be more exposed to danger in extremely remote or highly trafficked areas.

Collaborative and swarming UAVs have also garnered interest for logistical tasks. The USMC recently tested miniature drones with a centralised "Hive" that, together, drop more than 400 sorties per day, made up of "empty magazines, canteens and MRE ration packs on demand." [46]. Whereas the standard cycle to fulfil this resupply task is usually 96 h, this reportedly reduces it to mere minutes [47]. Such "Hives" may prove useful for last-mile logistics, particularly across challenging terrain. If systematised in the future, this may facilitate rapid deployments and sustainment. Beyond reconnaissance and attack roles, unmanned systems—on their own or in swarms—help move away from supply-based logistics by offering more, diverse opportunities to deliver logistics packages instead of items piling up in warehouses.

3.3 Agile Infrastructure

In a military logistics context, agility refers to the relevancy, support tailoring, and flexibility of systems to sustain performance over time [48]. The term infrastructure refers not only to the physical and IT systems, but also to the ability of units scaling up or down and of individuals training and reacting in dynamic environments where facts on the ground may change rapidly [5]. Borrowing from these definitions, various emerging technologies may be leveraged today to compensate for manpower, increase the modularity of logistical units, and using leaner deployments to project power more rapidly.

Just as the NATO Combined Joint Task Forces initiative envisioned modular "Lego Legions"—the design and deployment of various "force packages tailored for specific missions" [49]—more than two decades prior, Lego logisticians could employ additive manufacturing, unmanned systems, green energy technologies, and augmented reality in order to lower the logistical footprint without sacrificing key logistics capabilities.

While the application of many emerging technologies represents incremental innovation, preliminary uses of additive manufacturing as portable production facilities suggest a more fundamental, transformative effect. Operational readiness can be enhanced by cutting down on wait times for delivery of spare parts and components and also by flattening supply chains that have proved difficult to monitor, thereby achieving visibility over logistics assets located far from production chains.

Although the US Navy's use of 3D printing, also known as additive manufacturing, for prototypes dates back to the early 1990s, it was only in 2016 that 3D-printed flight-critical parts were added to the arsenal. On 29 July 2016, US Navy MV-22B *Osprey* aircraft successfully completed test flights a mere 18 months after the Naval Air Systems Command (NAVAIR) issued its additive manufacturing roadmap that included the benchmark of printing flight-critical parts within three years [40]. It is also notable that US Navy ships are designed to include 3D printers, with the USS *Harry S Truman* and USS *Kearsarge* following the USS *Essex* in suit as the first vessel equipped with a 3D printer. Rather than connecting to the land-based supply chain, the USS *Essex* printed its own oil reservoir caps, deck drain covers, and some medical supplies. The US Marine Corps has also been at the forefront of 3D-printing spare parts, notably with a Deputy Director at the Next Generation Logistics (NexLog) headquarters dedicated to additive manufacturing [50].

When transportation costs, particularly across domains, are more expensive than production or storage, the sophistication of the 3D-printed or-scanned parts becomes secondary. To an extent it does not matter whether the 3D-printed and-scanned components are complex or the more benign "nuts and bolts" critical to operations. In the latter category are TruClips, clasps used on handheld radios invented by US Navy soldiers. Ultimately the main cost is transportation, particularly across domains. Because additive manufacturing minimises transportation for the array of parts and components, it significantly decreases the cost. The TruClip, for example, costs six cents whereas the previous requisitions—from production to delivery and storage—cost $615 to replace [51]. The TruClip example also highlights how additive manufacturing enables creativity, whereby more "novel design concepts" can be entertained when initial design constraints fall to the wayside [40]. This form of creative thinking responds to calls for greater mental agility [5].

Other applications of additive manufacturing are possible, but not necessarily desirable or advantageous. For instance, in August 2018, the USMC 3D-printed barracks [8]. In forward operating bases, however, this solution may be too bulky. Furthermore, if taking into account the transportation for the specialised raw materials which are less likely to be available in the local supply chain than traditional construction materials. This example brings to mind the adage: if it isn't broken, don't fix it. With the adoption of emerging technologies in mind, agility should also mean sufficient critical thinking over whether solutions are overengineered.

Green energy technologies have also been deployed in order to enhance agility and manoeuvrability of forces, shortening logistical tails in the process [52]. DoD has long looked into biofuels and other alternative sources that generate energy to untether troops from insatiable fossil fuel demand to enhance troop safety, seek greater energy independence, and generate cost savings on fuel [53, 54]. As mentioned previously, frequent attacks on supply routes in Iraq and Afghanistan led DoD to prioritise protection of fuel and water resupply convoys.

Alternative energy sources, such as biofuels, have proven controversial with the US Navy. In 2012, biofuels used by the US Navy cost $27 per gallon as opposed to \$3.50 for conventional military fuels. Having received congressional attention, subsequent restrictions on biofuels have made it more difficult for increased usage [55]. Alternatives to batteries, however, have gained significant traction as a tool used to lighten soldiers' loads. Energy storage technologies such as fuel cells, which the US Army has tested for usage in its vehicles [56], also exist in portable varieties to replace heavy chargers and batteries with short battery life [43, 57]. Because they can be used in dismounted operations, the logistical load is physically lighter to transport and also allows troops to be more self-sufficient, requiring a smaller logistical footprint. Fuel cells are not a silver bullet, particularly due to their high operating temperatures and flammability. Nonetheless, certain portable fuel cells have undergone ruggedization tests with the US Army, with follow-on orders for special operations units suggesting positive results [58–60].

The unmanned and swarming developments outlined above may also serve to increase the agility of logisticians. While some unmanned or "uninhabited" systems act as one-for-one substitutions for manned platforms, other concepts including swarming may increase the agility of logisticians with new capabilities. Additional unmanned systems examples include the development of automated air-to-air refuelling solutions to reduce the workload and associated risk of air refuelling operators (AROs), who are in short supply and whose demand is expected to increase to fulfil future UAV refuelling needs [61].

Given the increasing digitisation of armed forces, agile infrastructure that permits greater connectivity will only become more important. Already many ground-based solutions exist, but aerial-based alternatives—in the form of cell-tower simulators and pseudo-satellites—may be attractive for the dismounted soldier. As seen in Ukraine, Russian-backed forces have been able to intimidate Ukrainian soldiers via threatening text message in areas without cell coverage. By using electronic warfare payloads aboard the *Orlan*-10 UAV, these Russian-backed forces transformed unmanned systems into cell-tower simulators [62].

Apart from the clear information and electronic warfare applications, such a technology could also be useful for logistics to enable greater manoeuvre for expeditionary forces operating away from forward operating bases. Pseudo-satellites, such as the *Zephyr*, which broke endurance records during its maiden flight of nearly 26 days, aim primarily at providing monitoring and surveillance capabilities [63]. In the future, other payloads may also make this surveillance device useful for satellite-simulated communications to, again, enable connectivity and dismount the soldier. Indeed the US Army, despite not having a line item for high-altitude UAVs, may consider solar-powered HAPS to serve as a future communications hub down the road [64] as operating environments become more contested.

Lastly, augmented reality can also complement repairmen in sophisticated systems, helping refocus the tooth-to-tail ratio. If augmented reality is more widely adopted, this could also mean that logisticians could afford to have a more generalist skillset, rather than needing specialists for each unique system. As demand for emerging technologies increases, this could help compensate for increasing maintenance of new systems. Indeed users in the commercial aerospace sector, such as Japan Airlines are already adopting augmented reality to close skill gaps [65]. Using video feed, be it 2D or 3D, is a force multiplier for maintenance specialists to allow them to be in many places at once and thus rebalance a key, shrinking asset: talent.

3.4 *Total Asset Visibility and Control*

Sensors—as discussed elsewhere in this chapter—are the backbone of total asset visibility control, whereby logisticians have precise, real-time control over sensor feeds to see where in the supply chain requisitions are located. Albeit to a lesser degree than DBL, total asset visibility was another concept from the late 1990s that was carried through logistics concepts and doctrine in the 2000s [7]. Automatic identification technologies (AITs) such as RFID and Item Unique Identification (IUID) sometimes looked like a panacea to track inventories, but have not proven infallible. Both emerging technologies and a conceptual overhaul of asset visibility are necessary to understand total asset visibility in today's context.

In theory, AITs offer manifold benefits: "better awareness of equipment location and status; faster, more accurate deliveries of needed supplies; and, of course, less manpower wasted on dreaded inventories." [66]. In theatre, the track record has been more complicated; implementation is time-intensive and the utility of RFID systems has not been exempt from interoperability issues—a perennial challenge for many of the IT systems used in logistics [7, 67, 68]. Furthermore it is difficult to track the added value of AITs because DoD has not been consistent at collecting key performance indicator or cost data related to implementation [69].

While AITs certainly serve a purpose, the presumption that total asset visibility is desirable in the first place has been called into question [70]. The sheer amount of data required for total asset visibility complicates data analysis. Information overload—or the phenomenon whereby "decision-makers face a level of information that is greater than their information processing capacity" [71]—can inhibit productivity. Applied to logistics, this means that total asset visibility may place the emphasis too strongly on *available* data rather than *actionable* data. As Kress argues, total asset visibility can make systems less robust [70].

Efficiency thus requires *proper* visibility before information collection produces diminishing marginal returns. Assuming logistics interventions—the kit of "parts, supplies, labor, special tools and equipment, and technical expertise," [26] as well as the IT systems that enable delivery—stay constant, increases in information asset visibility will produce only marginal improvements in the logistics pipeline. Having "a complete, perfect, and total" operational picture means overwhelming information technology systems and overcomplicating operational decisions—at both an economic and operational price [70]. In fact, in a 2017 GAO study on asset visibility, the term "total asset visibility" appears to be completely dropped [13].

Although 'snowballing supply' chain insecurities are not all logistics-centric, proper asset control should also mean mitigating supply chain deficiencies. The deficiencies are expansive: the globalised supply chain complicates this task, particularly for low-tier suppliers. For radar systems in US missile systems alone, the supply chain is at risk with "over 4,000 diminishing manufacturing sources and material shortages." [72]. Another example comes from the semiconductor industry: half of production is undertaken by non-US manufacturers—and a significant portion of the remaining half hail from offshore foundries that are more difficult to track [73]. Modernised supply chain management systems—enabled by sensors and blockchain—may minimise some of these tracking risks. However, more comprehensive recommendations are likely to come from the White House Office of Science and Technology Policy with strong inter-agency elements beyond the scope of this chapter.

Proper asset visibility, relying on snapshots and time-series being *good enough* rather than perfect, will have to take advantage of the myriad sensors and software systems used for inventory management. This is one area of logistics where transformation has a strong basis: the main management system employed by the US Army for asset visibility, the concept of which was already a few years old when mentioned in the 1999 *Army Logistician* special issue, is the Global Combat Support System-Army (GCSS). Launched in 2007 and fully implemented in 2015, it is one of the largest enterprise resource planning (ERP) systems in DoD [74, 75]. GCSS data has allowed the Army to cut $4 billion worth of spare parts it no longer needs and integrates various systems used across the 50 domestic depots [76].

In order to keep up with cutting-edge innovation, GCSS will need to integrate data collected from various types of sensors. In 2018, IoT was added to the list. Under an RFID contract, DLA acquired visibility solutions for 24,000 assets that use IoT sensors with a battery life of five to ten years, and which derive solar energy [77]. As mentioned above, despite their promise, IoT sensors may also prove too vulnerable to interference or tracking from unauthorised users.

Alternatively, blockchain may provide a more secure way forward if GCSS-Army is agile enough to leverage it. The decentralised ledger offered by blockchain may also help in logistical tasks, including validating transactions with fewer counterfeit parts and identifying alarming outliers from patterns in AIT. Blockchain may also become a useful antidote to cybersecurity concerns because of its "fast, secure, transparent" verification systems, stronger than encryption and other techniques currently used [78]. By always assuming the presence of an adversary on the network, no single node (or "block") can overpower a centralised command structure. Overall, when capitalising on any of these technologies for proper asset visibility control, practicality is a more useful parameter than is perfection [7].

3.5 Rapid Force Projection

Rapid force projection refers primarily to the deployment of forces from anywhere in the world within a given time period to sustain a truly global force—even over lines of communication beyond 10,000 miles. Deployment timelines may differ per the mission, in line with the concept of modular units discussed elsewhere in this paper. In 2001, the US Army stated its parameters for power projection: "deployability that will enable us to place a combat-capable brigade anywhere in the world in 96 h; put a division on the ground in 120 h; and five divisions on the ground in 30 days." [79]. The feasibility of this type of rapid deployment—including the logistics that would enable it—has been called into question, mainly because the predictability of deployments to Iraq and Afghanistan have taken precedence over the maintenance of rapid deployment capabilities [80, 81].

In this sense, it is the most platform-centric of the six pillars constitutive of an RML. Airlift and sealift, as well as amphibious capabilities, are the centrepiece of rapid initial entry, strike, and sustained manoeuvre forces. There are nonetheless smaller technological force multipliers to facilitate these phases of rapid force projection.

In addition to the very *long* distances, delivering "last mile logistics" over very *short* distances is often the most expensive, arduous part of the process. If strategic lift capabilities only get forces 99% of the way there, power projection can only truly be executed when the logisticians deliver that final, localised 1%. In Afghanistan, for instance, in active periods of fighting, about 100 vehicles are lost per week [82]. Fuel and water resupply convoys and supply routes have become frequent targets, increasing last-mile troop safety risks. Owing to safety concerns, *Chinook* helicopters have also been used to transport US embassy personnel over distances of 100 yards [83]. Be it to overcome geographical or policy-related reasons, capabilities insisting upon agility—such as autonomous last mile resupply capabilities mentioned above—must equally be considered for power projection and force sustainment.

The strain this poses on logistical forces is compounded by the extent to which military logistics is privatised [3]. With safety concerns on current supply routes, as well as the possibility of future battlespaces further expanding the frontlines and disintegrate identifiable safe areas [3, 84], it may become harder to recruit contractors willing to take on larger risk in the future. Concerns of troop and contractor safety may reinforce interest in unmanned or remotely controlled systems, particularly over the last mile.

Logistically, rapid force projection and sustainment will also mean choosing the best mode of transport for required materiel. Selecting aerial, land-based, or sea transportation dictates the speed, agility, and cost-effectiveness to deliver requisitions to units. Data analytics can be used as one way to improve this decision-making. The LOGSA partnership with Watson mentioned above also aims to optimise transportation requests and show when less expensive alternatives to air travel make sense. The aim is to have Watson audit 100% of transportation requests—and LOGSA anticipates that even at 10% capacity, such optimisation procedures would generate up to $100 million in cost savings per year [85].

Whereas insufficient location information during the First Gulf War led to spare part deliveries to inaccurate destinations, technology-enabled solutions may avoid such issues in uncontested environments. Additive manufacturing can also ensure that materiel—be they small platforms, components, spare parts, deployable infrastructure, medical supplies, or otherwise—is delivered to the correct locations. Taking advantage of "last mile" logistics solutions, the fact that 3D printers are being incorporated into designs for future ships suggest that sea-based logistics will become more important. Prepositioned ships equipped with 3D printers and associated materials would shift production lines from land to sea, making them more portable. This would be healthy for sustainment of naval and amphibious operations alike. Because flying items from the land-based supply chain out to sea is a costly endeavour, inversing the process to use sea-based production lines on land should not necessarily be the first choice.

Looking at the Indo-Pacific in particular, the combined risk of navy logistics concentrated in few, disparate locations and anti-access area denial threats, any capabilities that either reduce the logistical tail or localise it aboard ships merits consideration. Although this chapter focuses mostly on land-based capabilities, sealift capabilities merit commendation as a force multiplier for power projection in amphibious, land-, and sea-based operations alike. Unmanned vessels such as *Sea Hunter* from the Defense Advanced Projects Research Agency provide a useful model: if logistics vessels are retrofitted or built with autonomous navigation systems, they could concentrate combat support capabilities [86]. Furthermore, that *Sea Hunter* completed its first autonomously navigated voyage from San Diego to Pearl Harbor shows the maturity of the capability [87]. Using autonomous features—such as navigation—does not necessarily mean follow-on vessels to *Sea Hunter* have to follow the "paradigm-bending" shift toward totally crewless platforms [88], but that they can rather employ autonomous features to refocus crew, including logisticians, on other core tasks. As a final note, partially autonomous systems are likely to be both less expensive than a fully autonomous platform and easier to integrate into existing doctrine and culture. In this way, game-changing autonomous systems can distribute authority to lower echelons and thus facilitate the goal of increasing modularity of units tailored for specific missions [89].

4 Principal-Agent Theory and Information Asymmetries in Logistics Transformation

A key consideration for enhanced commercial technology utilisation in agile logistics is whether military operators are able to adapt or change the equipment. As noted by Erbel and Kinsey, that the private-sector owner maintains intellectual property rights (IPR) and technical data means operators, even in theatre, may not have the permission or assistance. Since the 1990s, this has corresponded in increased contractors operating in theatre [3].

Particularly as militaries consign logistical know-how to contractors, in-house knowledge of digitisation and machine learning may further degrade. Any "Amazonisation" of military logistics will have to take this into account and consider two scenarios. First is that the private sector will leverage ownership of the data to maximise earnings. Second is the possibility that the armed forces will lose out on these capabilities when private-sector owners are unwilling to share the data and analytics.

The first scenario is clearly illustrated by anticipatory logistics analytics. On the one hand, it is relatively simple to adapt predictive maintenance software from commercial technologies to work for the right types of sensor and training data needed by military users [90]. On the other hand, however, because the diagnostic information is owned by private sector operators, it is considered a separate capability that will cost dear. As summarised by the head of Air Mobility Command: "[o]ur airplanes have the capability right now. [...] All I have to do is put it on contract. We have the capability; we just didn't buy the capability." [91]. The same could be true of numerous technologies and analytical skills described above, ranging from augmented reality to upgraded software for unmanned systems to digital twins. In other words, the principal-agent theory animates logistics transformation by rendering logisticians and acquisition professionals reliant on the data owners and private-sector operators.

The second scenario is increasingly plausible as many of the cutting-edge companies in tech corridors perceive a cultural gulf with and distaste for the military. Whereas moral hazard used to refer to private contractors having an incentive to elongate military activity and their profit at odds with the missions at hand, the pendulum may have swung with high-tech companies reluctant to profit from work they perceive as immoral. Leading high-tech companies that own the data-analysis tools—not the data itself—may not be willing to share it with the Department of Defense. This was the case in June 2018 when Google announced it would not renew its *Project Maven* contract: in response to internal pressure from enraged employees, Google stepped away from using its AI-driven analytics to process UAV imagery. In the words of former Deputy Secretary of Defense Robert Work, "Google created a big moral hazard for itself by saying it doesn't want to use any of its AI technology to take human life. But they didn't say anything about the lives that could be saved." [92]. If logistics transformation means taking advantage of game-changing technologies and analytical tools, then narrowing this cultural gulf will only rise in importance.

That said, while worrisome to DoD about the trajectory of collaboration with commercial technology operators, the *Project Maven* controversy is distinct from military logistics-related tasks, which generally stay out of the limelight in comparison to more taboo issues, such as drone strikes. Many commercial-technology companies continue to cooperate with DoD, which also has ongoing efforts to improve relations with Silicon Valley and other tech corridors in the United States. Nonetheless, there is danger that an insidious pattern will emerge with negative effects on sharing data and analytics with the military. As such, both willingness and ability are key to military logistics transformation.

5 Conclusion: Adequacy of the Logistics Footprint

When considering the technological advancements above, two factors should ground analysis in reality. Firstly, the creation of a central "nervous system" of systems [93] depends overwhelmingly on strong connectivity, integration, and interoperability. If the shift toward network-centric warfare is at all informative, this process is far more difficult to achieve than presumed. Some predicted, or even feared, that advances in computing power would render war into a series of targeting exercises. The past two decades—characterised by interoperability issues, difficulty adjusting to shortened technology lifecycles, mounting cybersecurity and spectrum management concerns, and underwhelming adoption of commercially developed technologies—show anything but this.

The future will judge some of these technological enablers as hyped gadgets rather than pragmatic solutions [94]. With regards to Iranian and Gulf militaries, the political analyst Stephen Cordesman referred to the "glitter factor"[2]—or the mesmerisation with technology rather than the organisational and operational constructs that allow militaries to leverage technologically enabled advantages—of military capability development. Similarly, logisticians must not succumb to the seductiveness or *glitter* of these emerging technologies. "Technological romanticism," or the tendency to see technology as a panacea for military innovation [97], can lead to an unhealthy overreliance on technologies for logistics transformation. Subservience to technologies would be just as counterintuitive to logistics transformation as is the scant technological adoption experienced today and could create vulnerabilities in more than one way.

Vulnerabilities can stem from the complexity of using emerging technologies. From a manpower perspective, specialised systems that require more training than typical plug-and-play solutions may pose a difficulty for insufficiently staffed combat-support elements. Technology lifecycles pose another challenge: with hardware turning over in less than 36 months and software technology cycles lasting mere weeks, government-developed technologies could already be obsolete by the time they are fielded [98].

Furthermore, when roles become too specialised, reliance on technologies can atrophy human critical thinking abilities—either because human cognition stems from a more useful logic than that offered by fallible technology, or because the technology is not working. In addition to the less-than-perfect track record of technologies working in battlefield environments, such as those described in the asset visibility section above, the future battlespace is more likely to be contested than those in which the US armed forces operate today. If resources are allocated to automation at the expense of training and without adequate changes to organisational culture, then symbiosis between man and machine will not be achieved.

---

[2]The term first appeared in 1998 in commentaries on acquisitions that prioritised high-tech considerations over investments in training, recovery, sustainability, and interoperability. In a vein similar to that discussed here, other publications have emphasised the "glitter factor" as a distraction from the requisite qualities for a revolution in military affairs. See: [95, 96].

Symbiosis between logisticians and machines also depends on building trust. On one end of the spectrum, lack of trust in systems will mean expensive investments with little output. In culturally conservative organisations such as militaries, aversion to change may mean logisticians find workarounds to make choices in their comfort zone or with their own logic rather than with new systems [15, 39]. That the F-35 ALIS has exemplified these exact issues [39] does not bode well for centralised logistics systems for seamless logistics systems for other equipment. On the other end, when systems work and seem credible, humans can develop a blind trust even without understanding the complexity of or the logic behind decisions machines make [99–101].

The space between these goalposts, however, is large. As exemplified by positive experiences with additive manufacturing and software tools like GCSS-Army and JIGSAW, the adoption of emerging technologies and commercially driven innovation can still be successful in change-averse organisations like militaries. Software that proves agile and interoperable can be helpful without having to close the information loop completely. Instead, the logic of diminishing marginal returns that informs the difference between total versus proper asset control could be applied to the other RML tenets. With regards to agility, for instance, 3D-printed barracks may be perceived as an overengineered response to a non-existent problem; similarly for seamlessness, total completion of a closed information loop may not be constructive.

The original concept of an "adequate logistics footprint" described in the 1999 special issue of *Army Logistician* referred to the sufficiency of combat support resources. While this sought to establish a *floor* to make sure that logistics would not be stretched too thin, current definitions of "adequacy," as informed by twenty years of implementing new IT systems and experimenting with other emerging technologies, could also enable sustainable transformation by considering a *ceiling*, or the equilibrium point past which returns are marginal and investments inefficient. Instead of seeking complete awareness over everything at all times, system design for better logistics should require making choices on data collection priorities, ratios of logisticians to robots, and saturation of 3D printers and green energy technologies across units.

As a field driven by data and accustomed to close coordination with private-sector operators and innovators, logistics may be a prime testbed for force transformation. When understood as an enabler of action at the strategic, operational, and tactical levels, logistics transformation is a sensible gateway toward forces using emerging technologies to project power and sustain forces with smaller, more modular units.

References

1. Reimer DJ (1999) A note from the chief of staff of the army on the revolution in military logistics. In: The revolution in military logistics, vol 31, no 1. Army Logistician
2. van Creveld M (1977) Supplying war: logistics from Wallenstein to Patton. Cambridge University Press, Cambridge, p 1
3. Erbel M, Kinsey C (2018) Think again—supplying war: reappraising military tactics and its centrality to strategy and war. J Strat Stud 31(4):522

4. Gansler JS (2011) Democracy's arsenal: creating a twenty-first-century defense industry. MIT Press, Cambridge, p 215
5. O'Konski MJ (1999) Revolution in military logistics: an overview. In: The revolution in military logistics, vol 31, no 1. Army Logistician, p 2
6. Tripp RS, Amouzegar MA, McGarvey RG, Bereit R, George D, Cornuet J (2006) Sense and respond logistics: integrating prediction, responsiveness, and control capabilities, vol 9. RAND Corporation, p 38
7. Maccagnan V (2005) Logistics transformation: restarting a stalled process. U.S. Army War College Strategic Studies Institute, p 7
8. Grieco KA (2018) The 2018 national defense strategy: continuity and competition. Strat Stud Q 12(2):3–4
9. Harper J (2018) New national defense strategy prioritizes high-tech equipment, acquisition reforms. National Defense
10. Parlier GH (2005) Transforming U.S. army logistics: a strategic supply chain approach for inventory management. Land Warfare Paper 54. Association of the United States Army Institute of Land Warfare, Arlington, VA, September 2005. https://www.ausa.org/sites/default/files/LWP-54-Transforming-US-Army-Logistics-A-Supply-Chain-Approach-for-Inventory-Management.pdf
11. Spenard TM (2005) Force protection for distribution based logistics in asymmetric warfare. US Army Command and General Staff College School of Advanced Military Studies, Fort Leavenworth, KS, May 26, 2005. https://apps.dtic.mil/dtic/tr/fulltext/u2/a436192.pdf
12. Geospatial World (2013) Revolution in military logistics: the arithmetics [sic] of logistics, Noida, India, February 12, 2013. https://www.geospatialworld.net/article/revolution-in-military-logistics-the-arithmetics-of-logistics/
13. Merritt ZD (2017) Defense logistics: improved performance measures and information needed for assessing asset visibility initiatives. Report to Congressional Committees, Government Accountability Office, GAO-17-183, March 2017. https://www.gao.gov/assets/690/683413.pdf
14. Biddle S (2004) Military power—explaining victory and defeat in modern battle. Princeton University Press, Princeton, p 133
15. Beaumont D (2018) What we need to be: building the defence logistician—part two. Logistics in War, Canberra, Australia, June 22, 2018. https://logisticsinwar.com/2018/06/22/what-we-need-to-be-building-the-defence-logistician-part-two/
16. Heivilin DM (1991) Desert shield/storm logistics: observations by U.S. military personnel. Report to the Chairman, Subcommittee on Oversight of Government Management, Committee on Governmental Affairs, U.S. Senate, Government Accountability Office, GAO/NSIAD-92-26, November 1991. https://www.gao.gov/assets/160/151248.pdf
17. Pagonis WG, Krause MD (1992) Operational logistics and the gulf war. Land Warfare Paper 54. Association of the United States Army Institute of Land Warfare, Arlington, VA, October 1992, p 13. https://apps.dtic.mil/dtic/tr/fulltext/u2/a278028.pdf
18. Heivilin DM (1991) Desert shield/storm logistics: observations by U.S. military personnel. Report to the Chairman, Subcommittee on Oversight of Government Management, Committee on Governmental Affairs, U.S. Senate, Government Accountability Office, GAO/NSIAD-92-26, November 1991, p 26. https://www.gao.gov/assets/160/151248.pdf
19. Lancaster J (1991) Logistical shortcomings of 'Desert Storm' cited. Washington Post, December 31, 1991. https://www.washingtonpost.com/archive/politics/1991/12/31/logistical-shortcomings-of-desert-storm-cited/249a2bfb-12f1-443d-856f-e36fe26e374c/
20. MacGregor DA. Setting the terms of future battle for force XXI. Association of the United States Army Institute of Land Warfare, June 20, 1995. https://www.ausa.org/sites/default/files/LWP-20-Setting-the-Terms-of-Future-Battle-for-Force-XXI.pdf
21. Mongilutz MJ (1997) Force XXI logistics: operational distribution on the future battlefield. Naval War College, February 7, 1997. http://www.dtic.mildtic/tr/fulltext/u2/a324811.pdf
22. McKay R, Flowers K (2000) Transformation in army logistics. Military Rev Prof J U.S. Army. U.S. Army Combined Arms Center, Fort Leavenworth, KS, September–October 2000, p 46

23. Whitaker A, Sarder MDB, Ali A (2011) A literary review of distribution based logistics in operation Iraqi freedom. In: Paper presented at the international conference on industrial engineering and operations management, Kuala Lumpur, Malaysia, January 22–24, 2011. pp 972–974
24. Corbett DA (2007) Logistics transformation: the paradigm shift. United States Army Command and General Staff College School of Advanced Military Studies, Fort Leavenworth, KS, May, 14 2007. pp 25–34
25. Bestercy R (2008) The hybrid supply chain model: commercial logistics support and performance-based logistics for the department of defense. J Contract Manag 97–108. staging.ncmahq.rd.net/docs/default-source/default-document-library/articles/jcm08---pages-97-108
26. Payne D (1999) Distribution-based logistics. In: The revolution in military logistics vol 31, no 1. Army Logistician, p 38
27. Berteau DJ (2015) Sustainment and logistics in better buying power. Defense AT&L, p 38
28. Hunter A, Sanders G, Ellman J, Howe A, McDonald G (2018) Use of incentives in performance-based logistics contracting. Center for Strategic and International Studies, April 2018. https://csis-prod.s3.amazonaws.com/s3fs-public/publication/180406_Sanders_IncentivesInPerformanceBasedLogistics.pdf?4TJkLBqvQJ5R_1ClRyPoabwvg75ttLo0
29. Wallace R, Hardy CJ (1999) Seamless logistics system. In: The revolution in military logistics, vol 31, no 1, Army Logistician, p 18. http://www.alu.army.mil/alog/1999/janfeb99/pdf/janfeb1999.pdf
30. Hlad J (2018) DIUx making big impact in middle east wars. Air Force Magazine, February 9, 2018. http://www.airforcemag.com/Features/Pages/2018/February%202018/DIUx-Making-Big-Impact-in-Middle-East-Wars.aspx
31. Hlad J (2018) Forward deployed: raven, jigsaw, and chainsaw, vol 101, no 4. Air Force Magazine, p 8. http://www.airforcemag.com/MagazineArchive/Magazine%20Documents/2018/April%202018/Air%20Force%20Magazine%20April%202018%20Full%20Issue.pdf
32. Mahamuni A (2018) Internet of things, machine learning, and artificial intelligence in the modern supply chain and transportation. Def Transp J 74(1):16
33. Russell C (2016) Bulk fuel: actions needed to improve DOD's fuel consumption budget data. Report to Congressional Committees, United States Government Accountability Office, September 2016, p 11. https://www.gao.gov/assets/680/679682.pdf
34. DoD policy recommendations for the internet of things (IoT). Chief Information Officer, Department of Defense, December 2016, p 5. https://www.hsdl.org/?view&did=799676
35. Shetty S (2017) How to use digital twins in your IoT strategy. Gartner, June 8, 2017. https://www.gartner.com/smarterwithgartner/how-to-use-digital-twins-in-your-iot-strategy/
36. West TD, Blackburn M (2017) Is digital thread/digital twin affordable? a systematic assessment of the cost of DoD's latest Manhattan project. Procedia Comput Sci 114:47–56
37. Kirschbaum JW, Mazanec BM (2018) National security: long-range emerging threats facing the United States as identified by federal agencies. Report to Congressional Committees, United States Government Accountability Office, GAO-19-204SP, December 18, 2018, p 8. https://www.gao.gov/assets/700/695981.pdf
38. Clark C (2016) Operational testers flag F-35 software issues. Breaking Defense, January 25, 2016. https://breakingdefense.com/2016/01/operational-testers-flag-f-35-software-issues/
39. Insinna V (2019) Bad data in F-35 logistics system resulting in lost missions. Defense News, February 1, 2019. https://www.defensenews.com/air/2019/02/01/bad-data-in-f-35-logistics-system-resulting-in-lost-missions/
40. Newman J (2016) Printing the future: additive manufacturing promises to boost naval aviation readiness, vol 98, no 4. Naval Aviation News, pp 26–31
41. Sayler KM (2019) Artificial intelligence and national security. Congressional Research Service, Report R45178-Version 4—Updated, Updated January 30, 2019. https://fas.org/sgp/crs/natsec/R45178.pdf
42. Weisgerber M (2017) For sale: artificial intelligence that teaches itself. Defense One, July 6, 2017. https://www.defenseone.com/technology/2017/07/sparkscognition-artificial-intelligence-military/139225/

43. Stanley-Lockman Z (2018) Enhancing operational and cost-effectiveness: utility of 'Green Defence' to small nation-states. S. Rajaratnam School of International Studies (Singapore), March 14, 2018. https://www.rsis.edu.sg/rsis-publication/idss/enhancing-operational-and-cost-effectiveness-utility-of-green-defence-to-small-nation-states/#.W1l7i9gzY6g
44. Apthorp C (2017) Using autonomy to supply 'the last mile'. Army Technology, June 25, 2017. https://www.army-technology.com/features/featureusing-autonomy-to-supply-the-last-mile-5852408/
45. Defence and Security Accelerator funded contracts: 1 April 2017 to 31 March 2018. Defence Science and Technology Laboratory, Defence and Security Accelerator, Government of the United Kingdom, updated June 25, 2019. https://www.gov.uk/government/publications/accelerator-funded-contracts/accelerator-funded-contracts-1-april-2017-to-31-march-2018#autonomous-last-mile-resupply
46. Master W (2018) Drone delivers direct to grunt—Marine Corps Hive UAV. Platform Unmanned Cargo Aircraft, Eindhoven, Netherlands, March 16, 2018. https://www.platformuca.org/drones/drone-delivers-direct-grunt-marine-corps-hive-uav/
47. Freedberg SJ Jr (2018) Drone delivery, direct to the grunt: marines experiment with Hive UAVs. Breaking Defense, March 14, 2018. https://breakingdefense.com/2018/03/drone-delivery-direct-to-the-grunt-marines-experiment-with-hive-uavs/
48. Taylor III WH, Fowler RT (1999) Achieving an agile defense infrastructure. In: The revolution in military logistics, vol 31, no 1. Army Logistician, p 63. http://www.alu.army.mil/alog/1999/janfeb99/pdf/janfeb1999.pdf
49. Sloan SR (1994) Combined joint task forces (CJTF) and new missions for NATO. Congressional Research Service, Report 94-249S, March 17, 1994. http://www.dtic.mil/dtic/tr/fulltext/u2/a338635.pdf
50. Judson J. Marine corps looks to 3-D printing to make spare parts downrange. Defense News, September 11, 2017. https://www.defensenews.com/smr/equipping-the-warfighter/2017/09/11/marine-corps-looks-to-3-d-printing-to-make-spare-parts-downrange/
51. Church C (2016) Truman sailors' 3D-invention to be used by astronauts on ISS. Stars and Stripes, June 2, 2016. https://www.stripes.com/truman-sailors-3-d-invention-to-be-used-by-astronauts-on-iss-1.412679
52. Schuetz T, Stanley-Lockman Z (2017) Smart logistics for future armed forces. European Union Institute for Security Studies, November 8, 2017. https://www.iss.europa.eu/content/smart-logistics-future-armed-forces
53. Eady DS, Siegel SB, Steven Bell R, Dicke SH (2009) Sustain the mission project: casualty factors for fuel and water resupply convoys. Army Environmental Policy Institute, September 17, 2009. https://apps.dtic.mil/dtic/tr/fulltext/u2/b356341.pdf
54. Tiffen A (2014) Going green on the battlefield saves lives. War On The Rocks, May 22, 2014. https://warontherocks.com/2014/05/going-green-on-the-battlefield-saves-lives/
55. Cardwell D (2012) Military spending on biofuels draws fire. New York Times, August 27, 2012. https://www.nytimes.com/2012/08/28/business/military-spending-on-biofuels-draws-fire.html
56. Aliotta J (2018) U.S. Army TARDEC demos ZH2 fuel cell vehicle at Schofield with 25th Infantry. US Army Tank Automotive Research, Development and Engineering Center, February 8, 2018. https://www.army.mil/article/200366/us_army_tardec_demos_zh2_fuel_cell_vehicle_at_schofield_with_25th_infantry>
57. Fuel cells and other emerging manportable power technologies for the NATO warfighter. Part 1: power sources for manportable/manwearable applications. STO Technical Report, Science and Technology Organization, North Atlantic Treaty Organization, October 2014. https://apps.dtic.mil/dtic/tr/fulltext/u2/a616465.pdf
58. Protonex, raytheon to optimize power systems for US army. Fuel Cells Bull 2007(11):8 (November 2007)
59. Protonex follow-on power manager order from US army. Fuel Cells Bull 2016(1):11 (January 2016)

60. SFC delivers next-gen fuel cell systems to international military. Fuel Cells Bull 2017(2):4 (February 2017)
61. Airbus performs world's first automatic air-to-air refuelling contact with large aircraft receiver. Airbus, July 12, 2018. https://www.airbus.com/newsroom/press-releases/en/2018/07/Airbus-performs-worlds-first-automatic-air-to-air-refuelling-contact-with-large-aircraft-receiver.html
62. Electronic warfare by Drone and SMS. DFRLab, May 18, 2017. https://medium.com/dfrlab/electronic-warfare-by-drone-and-sms-7fec6aa7d696
63. Airbus Zephyr solar high altitude pseudo-satellite flies for longer than any other aircraft during its successful maiden flight. Airbus, August 8, 2018. https://www.airbus.com/newsroom/press-releases/en/2018/08/Airbus-Zephyr-Solar-High-Altitude-Pseudo-Satellite-flies-for-longer-than-any-other-aircraft.html
64. Reim G (2018) US Army eyes high-altitude UAVs like Airbus Zephyr. Flight Global, October 10, 2018. https://www.flightglobal.com/news/articles/us-army-eyes-high-altitude-uavs-like-airbus-zephyr-452551/
65. Canaday H (2017) Augmented reality could help close skill gap in maintenance. Aviation Week, March 9 2017. http://aviationweek.com/aviation-maintenance-and-support-software/augmented-reality-could-help-close-skill-gap-maintenance
66. Mariani J, Williams B, Loubert B (2015) Continuing the march: the past, present, and future of the IoT in the military. Deloitte, August 6, 2015. https://www2.deloitte.com/insights/us/en/focus/internet-of-things/iot-in-military-defense-industry.html
67. Edwards JE (2011) Defense logistics: DOD needs to take additional actions to address challenges in supply chain management. United States Government Accountability Office, Report to Congressional Committees, GAO-11-569, July 2011. https://www.gao.gov/assets/330/322061.pdf
68. Merritt ZD (2012) Defense logistics: improvements needed to enhance DOD's management approach and implementation of item unique identification technology. Report to the Subcommittee on Readiness, Committee on Armed Services, House of Representatives, United States Government Accountability Office, GAO-12-482, May 2012. https://www.gao.gov/assets/600/590619.pdf
69. Solis WM (2009) Defense logistics: lack of key information may impede DOD's ability to improve supply chain management. Report to Congressional Committees, United States Government Accountability Office, Report GAO-09-150, January 2009, p 7. https://apps.dtic.mil/dtic/tr/fulltext/u2/a492515.pdf
70. Kress M (2016) Operational logistics: the art and science of sustaining military operations, 2nd edn. Springer International Publishing, Basel, pp 101–102
71. Roetzler PG (2018) Information overload in the information age: a review of the literature from business administration, business psychology, and related disciplines with a bibliometric approach and framework development. Business Research, Goettingen, Germany, July 2018. p 2
72. Assessing and strengthening the manufacturing and defense industrial base and supply chain resiliency of the United States: report to President Donald J. Trump by the interagency task force in fulfillment of executive order 13806. US Department of Defense, September 2018. https://s3.amazonaws.com/static.militarytimes.com/assets/eo-13806-report-final.pdf
73. Radack DJ, Cohen BS, Leheny RF, Sharma V, Slusarczuk MMG (2016) Semiconductor industrial base focus study—final report. Institute for Defense Analyses, Alexandria, Virginia, December 2016, pp 1–3. https://www.ida.org/research-and-publications/publications/all/s/se/semiconductor-industrial-base-focus-study-final-report
74. Kenyon H (2011) Army readies major overhaul of logistics support. GCN Media, Virginia, USA, September 22, 2011. https://gcn.com/articles/2011/09/22/army-new-enterprise-logistics-support-system.aspx
75. Deaton K (2016) GCSS-Army: the future of army logistics. United States Army, November 1, 2016. https://www.army.mil/article/176893/gcss_army_the_future_of_army_logistics

76. Serbu J (2016) Army says logistics IT system has saved at least $6 billion. Federal News Network, July 25, 2016. https://federalnewsnetwork.com/dod-reporters-notebook-jared-serbu/2016/07/army-says-logistics-system-saved-least-6-billion/
77. Swedberg C (2018) DLA plans IoT solution for tracking assets at Texas facility. RFID J. https://www.rfidjournal.com/articles/view?18056
78. Marconi JM (2017) What are the business prospects for blockchain and big data (part 1 of 3). National Defense Transportation Association, September 13, 2017. https://www.ndtahq.com/blockchain-and-big-data/
79. Vick A, Orletsky D, Pirnie B, Jones S (2002) The Stryker brigade combat team rethinking strategic responsiveness and assessing deployment options. RAND Corporation, p 79. https://www.rand.org/content/dam/rand/pubs/monograph_reports/2005/MR1606.pdf
80. Trubenbach J (2014) Reinvigorating the army's deployment readiness. Army Sustainment, p 22. www.alu.army.mil/alog/2014/JulAug14/PDF/128786.pdf
81. Freedberg SJ Jr (2016) Over where? Army struggles to relearn rapid deployment. Breaking Defense, March 16, 2016. https://breakingdefense.com/2016/03/over-where-army-struggles-to-relearn-rapid-deployment/
82. Gouré D (2018) The key to success in Afghanistan is logistics. Real Clear Defense, February 23, 2018. https://www.realcleardefense.com/articles/2018/02/23/the_key_to_success_in_afghanistan_is_logistics_113099.html
83. Nordland R (2017) U.S. expands kabul security zone, digging in for next decade. New York Times, September 16, 2017. https://www.nytimes.com/2017/09/16/world/asia/kabul-green-zone-afghanistan.html
84. Freedberg SJ Jr (2017) No safe place in next war: the army's expanded battlefield. Breaking Defense, September 22, 2017. https://breakingdefense.com/2017/09/no-safe-place-in-next-war-the-armys-expanded-battlefield/
85. Stone A (2017) Army logistics integrating new AI, cloud capabilities. C4ISR.net, September 7, 2017. https://www.c4isrnet.com/home/2017/09/07/army-logistics-integrating-new-ai-cloud-capabilities/
86. Wakim E (2019) Sealift is America's achilles heel in the age of great power competition. War On The Rocks, January 18, 2019. https://warontherocks.com/2019/01/sealift-is-americas-achilles-heel-in-the-age-of-great-power-competition/
87. Sea hunter reaches new milestone for autonomy. Leidos, Reston, VA, January 31, 2019. https://www.leidos.com/insights/sea-hunter-reaches-new-milestone-autonomy
88. Scharre P (2018) Army of none: autonomous weapons and the future of war. W.W. Norton & Company, New York, p 78
89. Popa CH et al (2018) Distributed maritime operations and unmanned systems tactical employment. Systems Engineering Capstone Report. Naval Postgraduate School, Monterey, CA, June 1, 2018. https://apps.dtic.mil/dtic/tr/fulltext/u2/1060065.pdf
90. Sayler KM (2019) Artificial intelligence and national security. Congressional Research Service, Report, R45178-Version 4-Updated, Updated January 30, 2019. https://fas.org/sgp/crs/natsec/R45178.pdf
91. Weisgerber M (2017) Defense firms to air force: want your planes' data? Pay up. Defense One, September 19, 2017. https://www.defenseone.com/technology/2017/09/military-planes-predictive-maintenance-technology/141133/
92. Ramm T, Harwell D (2018) Google CEO quietly met with military leaders at the Pentagon, seeking to smooth tensions over drone AI. Washington Post, October 5, 2018. https://www.washingtonpost.com/technology/2018/10/05/google-ceo-met-with-military-leaders-pentagon-seeking-smooth-tensions-over-drone-ai/
93. Tucker P (2017) The future the US military is constructing: a giant, armed nervous system. Defense One, September 26, 2017. https://www.defenseone.com/technology/2017/09/future-us-military-constructing-giant-armed-nervous-system/141303/
94. Panetta K (2018) 5 Trends emerge in the gartner hype cycle for emerging technologies 2018. Gartner, August 16, 2018. https://www.gartner.com/smarterwithgartner/5-trends-emerge-in-gartner-hype-cycle-for-emerging-technologies-2018/

95. Cordesman AH (1998) Military balance in the middle east—XIII: the southern gulf: Saudi Arabia, the UAE, and Yemen. Center for Strategic and International Studies, December 28, 1998, p 27. https://csis-prod.s3.amazonaws.com/s3fs-public/legacy_files/files/media/csis/pubs/mbmexiiisgulf122898%5B1%5D.pdf
96. Cordesman AH (1999) Iran's military forces in transition: conventional threats and weapons of mass destruction. Praeger, Westport, p 169
97. Adamsky D (2010) The culture of military innovation: the impact of cultural factors on the revolution in military affairs in Russia, the US, and Israel. Stanford University Press, Palo Alto, pp 85–87
98. Spataro NM (2018) FY2018 NIAC consolidated report: lean startup approach to capability development: evolving small batch innovation for dual-use technologies. Naval Innovation Advisory Council, August 1, 2018, p 3
99. Leetaru K (2016) In machines we trust: algorithms are getting too complex to understand. Forbes, January 4, 2016. https://www.forbes.com/sites/kalevleetaru/2016/01/04/in-machines-we-trust-algorithms-are-getting-too-complex-to-understand/#6efc83da33a5
100. Fry H (2018) Don't believe the algorithm. Wall Street J, September 5, 2018. https://www.wsj.com/articles/dont-believe-the-algorithm-1536157620
101. Scharre P (2018) Army of none: autonomous weapons and the future of war. W.W. Norton & Company, New York

The manufacturer's authorised representative in the EU is Springer Nature Customer Service Centre GmbH, Europaplatz 3, 69115 Heidelberg, Germany. If you have any concerns regarding our products, please contact ProductSafety@springernature.com

Printed and bound by CPI Group (UK) Ltd, Croydon, CR0 4YY
25/09/2026
02222892-0005